GREAT MINDS DON'T THINK ALIKE

GREAT MINDS DON'T THINK ALIKE

DEBATES ON CONSCIOUSNESS, REALITY, INTELLIGENCE, FAITH, TIME, AI, IMMORTALITY, AND THE HUMAN

EDITED AND WITH COMMENTARY BY
MARCELO GLEISER

Columbia University Press
New York

Columbia University Press

Publishers Since 1893

New York   Chichester, West Sussex

cup.columbia.edu

Copyright © 2022 Marcelo Gleiser

Library of Congress Cataloging-in-Publication Data
Names: Gleiser, Marcelo, author.
Title: Great minds don't think alike : debates on consciousness, reality, intelligence, faith, time, AI, immortality, and the human / by Marcelo Gleiser.
Description: New York : Columbia University Press, [2021] | Includes index.
Identifiers: LCCN 2021022050 (print) | LCCN 2021022051 (ebook) |
ISBN 9780231204101 (hardback) | ISBN 9780231204118 (trade paperback) |
ISBN 9780231555371 (ebook)
Subjects: LCSH: Philosophy.
Classification: LCC B53 .G49 2021 (print) | LCC B53 (ebook) | DDC 100—dc23
LC record available at https://lccn.loc.gov/2021022050
LC ebook record available at https://lccn.loc.gov/2021022051

Cover design: Noah Arlow

CONTENTS

INTRODUCTION

I n Fall 2016, I joined the stage of 92nd Street Y in New York City with neuroscientist Antonio Damasio and philosopher David Chalmers for a conversation on the "Mystery of Consciousness." This was the first in a series of public dialogues that I conducted for the next five years in theaters and universities across the United States. They were part of the activities of the Institute for Cross-Disciplinary Engagement at Dartmouth, which I founded with generous funding from the John Templeton Foundation. Our mission was to bring scientists and humanists together, in what I call constructive engagement, to discuss and confront some of the most challenging questions of our times, from the more abstract "What is the nature of reality?" to the more practical "What is the future of humanity in the age of AI?" Our motivation was the essential realization that such questions are too complex to be addressed one-dimensionally, either only by the sciences or only by the humanities. As with many questions that define our time, they call for a pluralistic approach, combining different ways of knowing, if we are to make progress in answering them. There are, of course, many questions that are very much within the sole province of either the sciences or the humanities and these, for obvious reasons, were not part of our dialogues. The selection of topics discussed is certainly not complete, but it will hopefully illustrate that the sciences and the humanities have much to say to one another in matters of great import and interest to our collective future.

This volume includes eight of these conversations, in some cases with questions from the audience as well. The topics are broad and timely, and the list of contributors is impressive, from Pulitzer and Templeton Prize winners to MacArthur "genius" grant fellows and well-known public intellectuals. We live in times when civil discourse is seriously threatened by bigotry and tribal entrenchment. My hope is that the conversations in this book will set an example for how people can engage in a fruitful exchange of ideas, even when there is disagreement.

BEYOND THE TWO CULTURE DIVIDE

"I believe the intellectual life of the whole of western society is increasingly being split into two polar groups. . . ." So wrote the British physicist and novelist C. P. Snow in his famous *The Two Cultures* Rede Lecture delivered at Cambridge University in 1959.1 Although Snow was mostly concerned with the divisions he felt in his own personal and professional experience between the "literary intellectuals" and "physical scientists," the two-culture split has come to symbolize a wider and ever-growing gulf in academia between the sciences and the humanities. This split, and the strife it often generates, is palpable in most universities, and it speaks directly to the heart of the liberal arts curriculum of schools across the globe, and to the markedly wrong, widespread perception that in a technology-driven world, the humanities are an anachronism.

The roots of this unfortunate split between the two cultures reach back beyond the Enlightenment and its discontents, having been amplified by an increasingly successful scientific enterprise and the consequent technologization of society. The seventeenth century marked a turning point in human intellectual history where what we now call the sciences started to carve their own path away from the Greek philosophical tradition. Kepler, Galileo, Descartes, Newton, Boyle, and many others took off as *natural philosophers*, concerned with the workings of nature as their Greek and Islamic forefathers had been, but now armed with a powerful new methodology whereby direct experimentation and data analysis were to describe a variety of terrestrial and celestial phenomena with mathematical precision. Their spectacular success changed the way we understand the cosmos and our place in it, creating, as a byproduct, a deep spiritual rift that has never been healed. If the human mind can understand the workings of the world without apparent limitations, what room, then, is there for mystery? For spiritual questioning? If the world is truly machine-like, operating under strict mathematical logic, what room then for doubt, for free will?

As influential thinkers promoted science as the sole source of truth, the humanities lost some of their clout, and the rift between the two cultures gained momentum. "Literary intellectuals at one pole—at the other scientists, and as the most representative, the physical scientists. Between the two a gulf of mutual incomprehension—sometimes (particularly among the young) hostility and dislike, but most of all lack of understanding," wrote Snow. Experts hid behind the jargons of their respective fields and either talked past each other, or worse, didn't talk to each other at all. The frontiers of knowledge broadened, academic departments multiplied, and with them, the walls separating experts in ever narrower subdisciplines.

Perhaps the greatest virtue of Snow's essay was to describe science as a culture. And that it surely is, both within its practices and practitioners and as a driver of profound changes in humanity's collective worldview since the seventeenth century. The relentless ascent of scientific thinking brought the contempt of many humanists who considered themselves as the only worthy intellectuals—scientists are technicians; humanists are intellectuals. Ensconced within their turf, most scientists returned the disdain, considering the humanities to be worthless for their intellectual pursuits. "Philosophy is useless," well-known scientists have proclaimed. "Religion is dead."

We can see the tension—and the issues it creates—most clearly in areas where science encroaches upon territory that has historically been chartered primarily by humanists. It is common to hear that science is about nature, while the humanities deal with values, virtue, morality, subjectivity, and aesthetics; hard to quantify concepts about which science has nothing or very little to say. To describe love as a set of biochemical reactions taking place due to the flow of a handful of neurotransmitters through certain regions of the brain is clearly important, but it does very little to describe what the experience of being in love feels like.

Such polarizations are deeply simplistic and are growing less relevant every day. Current developments in the physical, biological, and neurosciences deem such narrow-minded antagonism and mutual exclusion as problematic and downright corrosive. It limits progress and inhibits creativity. Many of the key issues of our times, an illustrative sample being the questions explored in this volume, call for a constructive engagement between the two cultures. It is our contention that the split between the sciences and the humanities is largely illusory and unnecessary, in need of a new integrative approach. We need to reach beyond traditional disciplinary boundaries to create truly cross-disciplinary ways of thinking. It is no longer enough to read Homer and Einstein or Milton and Newton as disjoint efforts to explore the complexities of the world and of human nature. The new mindset proposes that the complexities of the world are an intrinsic aspect of human nature as we experience reality. We cannot separate ourselves from a world that we are a part of. Any description or representation, any feeling or interpretation, is a manifestation of this embedding. Who we are and what we are form an irreducible whole.

The questions that call for an engagement between the sciences and the humanities are not restricted to academia. Consider the future of humanity in an ailing planet as we move toward a more thorough hybridization with machines. We currently extend our physical existence in space and time through our cell phones, while many scientists and humanists consider futuristic scenarios where we will transcend the body, becoming part human and part machine, with some even speculating that a singularity point will be reached when machines will become smarter than we are—although they are rather vague on the meaning of smarter. Such technological advances call into question the wisdom of our scientific advances, raising issues related to machine control, the ethics of manipulating humans and other lifeforms, the impact of robot-ization and artificial intelligence in the job market and in society, and our predatory relationship with our home planet. There is a new culture emerging, inspired by questions old and new that reside at the very core of our pursuit of knowledge. The choices we make now, as we shape our curricula and create academic departments and institutes and engage in discussions with the general public, will shape generations and the nature of intellectual cooperation for decades to come.

NOTE

1. C. P. Snow, *The Two Cultures and the Scientific Revolution*, (Cambridge: Cambridge University Press, 1959).

GREAT MINDS DON'T THINK ALIKE

1

THE MYSTERY OF CONSCIOUSNESS

A Dialogue Between a Neuroscientist and a Philosopher

DAVID CHALMERS AND ANTONIO DAMASIO

This conversation features two eminent modern thinkers who have never appeared together. **Antonio Damasio** is University Professor, David Dornsife Chair in Neuroscience, and Professor of Psychology, Philosophy, and Neurology at the University of Southern California, where he also directs the Brain and Creativity Institute. He has made seminal contributions to the understanding of brain processes, underlying emotions, feelings, decision-making, and consciousness. He has authored four books, most recently *Strange Order of Things: Life, Feeling and the Making of Cultures*. And he has an astounding 183,100 citations for his academic work. **David Chalmers** is a university professor of philosophy and neural science and co-director of the Center for Mind, Brain, and Consciousness at New York University. He also has an appointment at the Australia National University, where he used to direct the Centre for Consciousness. He works on the philosophy of mind and language, and on artificial intelligence and cognitive science. He's well known for his formulation of the hard problem of consciousness, which is the problem of explaining how and why we have first-person experiences. Essentially, we could phrase it like this: Why does the feeling that accompanies the awareness of sensory information exist at all? He's the author of *The Conscious Mind* and, more recently, *The*

Character of Consciousness, as well as many philosophical essays (I have to add that he's also the lead singer of the band Zombie Blues).

To contextualize our topic, here's a quote from a scientist:

> The passage from the physics of the brain to the corresponding facts of consciousness is unthinkable. Granted that a definite thought and a definite molecular action in the brain occurs simultaneously, we do not possess the intellectual organ, nor apparently any rudiment of the organ, which would enable us to pass by a process of reasoning from the one phenomenon to the other. They appear together, but we do not know why.
>
> Were our minds and senses so expanded, strengthened, and illuminated as to enable us to see and feel the very molecules of the brain, were we capable of following all their motions, all their groupings, all their electric discharges, if such there be, and were we intimately acquainted with the corresponding states of thought and feeling, we should be as far as ever from the solution of the problem. How are these physical processes connected with the facts of consciousness? The chasm between the two classes of phenomena would still remain intellectually impassable.
>
> Let the consciousness of love, for example, be associated with a right-handed spiral motion of the molecules of the brain and the consciousness of hate with a left-handed spiral motion. We should then know when we love, that the motion is in one direction and when we hate, that the motion is in the other. But the why would remain as unanswerable as before.

This text was written in 1868 by Victorian physicist John Tyndall when he addressed the physical section of the British Association of the Advancement of Science. I find it quite brilliant and remarkably prescient, although no doubt a lot has happened since. We are about to find out what, and I can't wait. What is consciousness? How can philosophy and neuroscience work together to help elucidate one of the most fascinating and mysterious questions we can ask? Or is consciousness so elusive that it remains outside human understanding?

A DAMASIO: I'd like to make a few comments at the beginning to introduce the problem, and I thought that it would be good to do so around the issue of nomenclature. That is because so very often, people have conversations about consciousness, and at a certain point one realizes that they are talking past each other—and that is because the word can mean so many things.

Let me start with a few notes about what consciousness is *not*. When you think about the word, all those of you who have a background in romance languages will immediately remember that the word used to mean *consciousness* is actually the same, with the same exact spelling, as the English word for *conscience*. And one thing that consciousness is not is a synonym of conscience.

Conscience is, of course, quite related. You cannot have conscience, in the proper sense, if you are not conscious. That would be an absurdity. But the fact remains that when Shakespeare has Hamlet say, "Thus conscience makes cowards of us all," he is not referring to consciousness. He is referring to the fact that Hamlet's contemplation of killing his uncle is not a good thing. It is something morally reproachable, and the moral reproach turns him into a coward. He will not be able to carry out that act.

Those of you who know Portuguese, for example, know that in Portuguese, when you mean consciousness, you use *consciência*, which is exactly the same word that you use if you're referring to conscience in the moral sense. A great big confusion. Exactly the same thing happens in French with *conscience* and in Italian with *coscienza*.

In English and in romance languages, the word for conscience came first. In English, a specific word for consciousness was coined much later, in the seventeenth century. In romance languages, an equivalent word was never coined. The lack of a word for consciousness is a source of confusion when one is talking across languages on a subject like this.

Another thing that consciousness is not is general awareness—for example, awareness of ideas. You will often come across sentences such as, "We now have consciousness of the gravity of the problem of global warming." Of course, to be aware of a problem requires that you're conscious, that you have acquired the appropriate knowledge and reasoned over that knowledge, but that does not correspond to the specific phenomenon of having a mental experience or having a subjective point of view.

Another thing that consciousness is not is complete wakefulness or vigilance. Of course, one cannot have mental experiences if one is not awake and aware and vigilant in the neurological sense, but one can actually be asleep and have a subjective point of view during the paradoxical consciousness of dreaming. When we dream, we do have a point of view; we do have experience; we can have feelingness (and then some)! But that is still different from regular, awake consciousness. The word paradoxical fits the dream situation.

Very often, in the world of medicine, a neurologist will talk about, for example, patients who have been involved in an accident and say that they "lost consciousness." Well, what the neurologist is referring to is not only the loss of mental experience—although this has been lost too—but rather the more fundamental loss of sensing and responding, conscious or otherwise. It is really like turning off the lights and not having the basic phenomena which support mental experiences, such as being awake and alert, as opposed to the precise removal of experience only.

This goes against the idea that consciousness and sentience are synonymous. Sentience is not a synonym of consciousness. Sentience is best used to describe an organism's ability to detect a stimulus and respond to that stimulus. Now, why is this distinction so important? For the following reason: unicellular organisms as simple as bacteria, which do not have a nucleus let alone a nervous system and a mind in the sense that we do, are certainly *sentient*. They can detect clearly and specifically certain stimuli and organize responses to them. So we need to distinguish sentience from consciousness. When we talk about having a mental experience, or consciousness, it is a precondition that we are awake and sentient. We can see wakefulness and sentience as stepping-stones to consciousness.

The phenomena that most interests us tonight, and I think that David would probably agree, is consciousness specifically understood as a mental experience that is felt and that has a point of view. The point of view happens to be that of the experiencing organism. At the moment, my mental experience includes the fact that I'm seeing David and Marcelo to my left, in a particular perspective. I have the feeling that goes with my being here, but I also have a perspective that is unique, my own point of view on the scene. And of course, David and Marcelo have different perspectives, along with the feelingness. That goes with holding a certain perspective with specific contents. Not only do I have a different perspective compared to theirs, but I have the automatic sense that their perspective is different, and that what I am looking at and what they are looking at is being looked at from different angles in space.

Discussions on consciousness have involved people in the arts and humanities, in philosophy of mind, neuroscience, psychology, and cognitive sciences of every very kind and flavor, and I think it's important to think about what it is that we need to do to advance the science of consciousness. This is something I know that David is interested in and it's one reason why it's good to have a dialogue with him. I believe we have very obvious points of agreement.

One important requirement for the success of this conversation is that we take phenomenology seriously. We very often end up hearing discussions where people are in apparent disagreement, and yet it's obvious that one person is thinking about looking at one's mind, being an observer of one's mind, while the other person is thinking about allegedly more objective things, such as, for example, the activity of a patch of neurons. There is no incompatibility. We need to be informed by both levels of analysis and we need to take seriously the fact that some phenomena are first-person phenomena that occur in our minds and that have to be observed, analyzed, described by one person, subjectively. There is a great tradition of taking phenomenology seriously and describing what is going on in the mind as accurately as possible, for example, in European philosophy, notably in Germany and in France.

If we can have phenomenological data collected from multiple persons in the same circumstances, there's no reason why the data should not be comparable with data on the activity of a patch of neurons. It's true that we cannot measure phenomena with the same kind of instruments that we use in a physiology experiment, but that doesn't mean that one should discount them, because if one does, then we will never have a science of consciousness. When David talks about a first-person perspective and a third-person perspective, he is saying the same thing with different words.

We also need to take biology seriously. Contrary to what we heard in the very nice speech of the gentleman physicist from the last century (that Marcelo read at the beginning), I think the days when we could afford to exclude data are over—whether it is phenomenological data, first-person observations, or data that comes from objective neural tissue experiments. For phenomena that can be studied from different perspectives, we need data from all of those perspectives. It doesn't make any sense to exclude data, because we're talking about something very complex that is not about neurons only, about brain systems only, about bodies and life only, and it's certainly not going to be only about something lost in the ether called *mind* or *psyche*. It may be that, in time, we will conclude that this integrative position is wrong, but right now, that is our best bet.

To take biology seriously means to take seriously more than just the neuro part of biology. To take biology seriously is to realize that there are other living organisms that we descend from. Bacteria, for example, are ancestors to all of us. The history of life began with unicellular organisms that did not have a nervous system and yet had absolutely remarkable properties in terms of the management of their life and complexity in the relations they held to other organisms. We need to place the issues of mind and consciousness in the perspective of this long trajectory that is at least 3.9 billion years old—billions of years.

We need to take into account that all living organisms are very complex creatures, even when they seem simple. And although many do not have nervous systems, they all actually have bodies. It is quite common to hear or read discussions on the nature of consciousness where it is assumed that consciousness is a product of the brain alone, that consciousness is produced by an isolated brain, completely forgetting about the body. The fact is that we do not have isolated brains and that most creatures on this planet do not have nervous systems at all, yet all have bodies capable of maintaining life intelligently and competently, even when nervous systems are missing.

As for humans, we have a nervous system with a very high level of central organization—the central nervous system, which includes sections such as the spinal cord, brain stem, a variety of ganglia at the base of the brain and the cerebral cortex. One problem with not taking this complex organization seriously is that discussions of consciousness, if they venture into biological territory, often talk only of the central nervous system and sometimes even only of the cerebral cortex, as if the rest of the nervous system was not dignified enough or complex enough to contribute importantly to anything like consciousness. In other words, not only do people forget the body, they end up forgetting most of the nervous system. This has got to be wrong. It cannot work that way.

If you ask me today, "Do you believe that minds and consciousness are created by brains alone?" my answer would be unequivocally, "No. I do not think that the brain alone is capable of doing it." So what else is needed? Well, we certainly need a full nervous system but also a nervous system that is inserted inside a body proper and fully interactive with it.

A mind is not the result of brains alone nor of nervous systems alone. It's the result of an integrated functionality that joins nervous systems and the body proper for the purpose of regulating life efficiently.

The partnership creates something that is very often missed—a complex system distributed within every nook and cranny of the organism and holding a dual relationship: neurons that go to every nook and cranny, and then neurons, or molecular signaling, that come from those nooks and crannies all over the body. This very complex arrangement allows for a huge amount of interaction.

The interaction began when the history of nervous systems began, in the Precambrian period, when there appeared simple nerve nets rather than full nervous systems, let alone brains. Such nets looked a bit like our reticular formation looks today, inside our spinal cords and brain stem. All of those nerve nets were affected by the chemical molecules that were circulating in the body.

So we have not only a possibility of interaction between body and brain, between body and nervous system at the level of the signaling that is neural and in which the body participates, but we also have a major interaction that happens chemically and that directly affects the nervous system but comes from the body. We need to realize that the nervous system is in fact surrounded by two orders of structure: one, the organism's interior, and the other, something that is in the surround of the organism, the outside world if you will. So it's not merely the nervous system looking at the outside world; it's the nervous system looking at the inside world and looking, at the same time, at the outside world.

In considering mental experiences and the subjective point of view, we need to address two more issues.

One is feeling or feelingness, something that has a direct connection to traditional conversations on consciousness and which is known in philosophy as *qualia*. For me this is a layered phenomenon. There's not a single kind of feelingness, but at least two main kinds. Spontaneous or homeostatic feelings mentally express the unfolding state of life within the organism and serve as either an alarm signal, such as pain, or as a signal that all is well, well-being. Provoked feelings mentally express our emotional reaction to the images that constantly appear in our minds. We react emotionally to these images with changes in physiology that prepare us to deal with the situations imaged—and this changed state of physiology in mentally expressed in feelings of joy or fear or admiration and so on, depending on the content of the provoking images.

Besides the feelingness that is the obligate accompaniment of all images, we must acknowledge that our mental images are oriented by a sort of a GPS that places them in relation to the perspective of our organism. Such images are constructed according to where in space our sensory portals are at any given moment. By sensory portals, I mean our eyes and ears and tactile corpuscles in the skin.

In conclusion, we are better off, for certain, than 148 years ago. Perhaps the key is to take an evolutionary perspective seriously. Unless we adopt that evolutionary perspective, things cannot possibly make sense, and will look far stranger than they need to be.

D CHALMERS: It's a privilege to talk about this topic with Antonio Damasio and Marcelo Gleiser. I think the mission of this institute to cross the disciplines, and in particular to bridge the sciences and the humanities, is an important one. And I think this topic of consciousness is really a very appropriate topic to be beginning with.

So the problem of consciousness is, I think, traditionally a problem at the very center of philosophy. How does the mind relate to the body? This is the traditional philosophical mind-body problem, which over time is turning into the consciousness-brain problem. How does consciousness relate to the brain? And how can we explain consciousness in terms of the brain, and perhaps the body?

It's gradually moving from being a philosophical problem to becoming a scientific problem, but I think it's not all the way there yet. It's still in the transitional stage. There are many very deep philosophical elements left to the problem.

They say that philosophy is the source of all the sciences. You know, physics started off as philosophy until they made it rigorous enough that people developed a methodology where they could perform experiments and agree on results. And then we branched the physicists off. The physics start-up has been really quite successful over the years. *[laughter]* We did the same for psychology in the nineteenth century and for linguistics and economics. And it's one of my dearest hopes that this will happen one of these days for the science of consciousness, and I think we're in the process of that happening. But there are still, I think, many very deep philosophical mysteries, which are very closely interacting with the science.

Antonio talked about definitions and I agree, it's very, very important to get clear on words. It's one of our favorite things to do as philosophers. And I should say for those of you who came expecting a knock-down, drag-out debate, I think I actually agreed with almost everything that Antonio said. Sorry to disappoint you. Nonetheless, maybe we have some different points of emphasis and we develop these thoughts in different directions.

On definitions, I'm happy with Antonio's definition of consciousness as mental experience or subjectivity. I quite like myself to put these two together and just define consciousness as subjective experience, the subjective experience of the mind and of the world by a subject from a first-person point of view. And for me, this includes everything from very simple experiences, like the experience of color—seeing the red of the exit sign, hearing my voice, feeling a pain—to very complex experiences, like the experience of very complex and rich emotion, the experience of complex philosophical thinking, of remembering a certain period when I was very young.

These are all elements of this grand movie of consciousness that we experience from a first-person point of view. What they all have in common is that they all feel like something. There's something it's like, as my colleague Tom Nagel at NYU has put it. There's something it's like to see, to feel pain, to feel emotions, to think, and that something it's like from the inside is for me the essence of consciousness. And I think for Antonio too.

I agree with the distinctions between consciousness and conscience and knowledge. I think it's not the same as those, or as wakefulness. I think there's a kind of consciousness in dreaming. Some people use the words awareness and sentience for the same phenomena. Other people use them differently. I don't want to get in too much of an argument about what those words mean, but I agree that if you understand sentience as stimulus and response, that's not what consciousness is. Consciousness is an experience.

This is what philosophers call phenomenal consciousness or experiential consciousness. I might distinguish it from two other things.

There's also self-consciousness, consciousness of the self. When I'm conscious, for example, of myself, that is a very familiar phenomenon for human beings. But much of the time we go through our lives conscious of the world, conscious of things around us without any focal consciousness of ourselves. And it may well be that self-consciousness is something that is evolutionarily more sophisticated and comes later than simple consciousness of the world. For example, it may be there are some organisms—fish, for example—that can feel a pain, can have a subjective experience of a pain, and would be conscious in this experiential sense. But whether fish are conscious of themselves, well, I don't know. There's a lot of arguments to be had about that. But it's very possible, as far as I'm concerned, that fish have the simple experiential consciousness without self-consciousness, which is something much more sophisticated.

I'd also distinguish between consciousness in the sense of subjective experience and consciousness in the sense of response, in the sense of being tied to certain kinds of behavior.

Psychologists and neuroscientists love to operationalize things. We can all talk about phenomena, such as, for example, my being conscious of a stimulus such that I'm able to point to it. I'm able to report it. I'm conscious of the exit sign. One operationalization of it is I can talk about it. I can say, "There's an exit sign there." That's a good sign of consciousness, but for me, that's not consciousness itself. The ability to respond, the ability to behave is not consciousness. Consciousness is the experience. And this kind of distinction between, say, experience and behavior or functioning really gets us, I think, to the core of the problem that consciousness poses in science.

Marcelo read this beautiful quote from Tyndall in the 1860s about the gap between any physical description of the brain and subjective experience, and I think this is a gap that confronts us today. I called it the hard problem of consciousness twenty years ago, but it was not a new insight. This is a hard problem. Tyndall quite clearly recognized that problem in 1868. Leibniz recognized it in 1714. It has a very long history.

I would say when it comes to explaining the phenomena of the mind, there are the phenomena of behavior—how we get around in the world, how we walk, how we talk, how we respond, and how we process information to make decisions. All those are very difficult problems, but we have a bead on how to explain them given the standard methods of science. Isolate a neural circuit; tell some computational story about the brain and the environment that explains ultimately how the brain produces this functioning, processes the information, giving rise to certain behavior. And that works very well for explaining certain phenomena, say of learning and memory and language use, even certain aspects, many aspects, of perception. But when it comes to the problem of consciousness, those are what we call the easy problems, precisely because we've got a bead on explaining them.

The hard problem is this problem of explaining subjective experience. How on earth does all this processing feel like something from the inside? There is this natural sense of a chasm here. Why should all this brain processing feel like anything at all? Why aren't we just like robots or zombies that have all this processing going on with no subjective experience at all?

The trouble is that the standard methods of science—which are so good for explaining things like behavior and information processing that produces behavior, which is terrific for explaining things like learning and memory and so on—seem to leave this problem untouched, because you can explain all that functioning and you'll still be left with the question of where does the subjective experience come in?

Speaking of mid-nineteenth century scientists, there was also Huxley, who said—just a few years before Tyndall, I think—the problem of how irritating brain tissue should give rise to subjective experience is as mysterious as how it is that the djinn appeared when Aladdin rubbed his lamp.

For many people, there's still a question of how any purely neurophysiological explanation of brain processes could tell you why there's consciousness there at all. Even bringing in the body or the environment, which I think is a very useful thing to do for many purposes, I think leaves this problem wide open. If I were to talk about this at length, I'd talk about some reasons why we might want to go beyond certain kinds of reductionism, where all this gets explained in terms of the physical world and bring in something new and fundamental to explain consciousness.

And my view is that we do need something new and fundamental to explain consciousness. But one way to think about this from a dialogue between philosophy and science is you might think, "OK, if you're going to be opposing reductionism, you're thereby opposing science." But this is not my view at all.

My view is actually that opposing an overly reductionist view of consciousness here actually opens the way to do science. I see the science of consciousness as an extremely important phenomenon that has flowered in all kinds of directions over the last twenty or twenty-five years, and Antonio's work has really been at the center of that—and any number of other people, as well, in neuroscience and psychology and philosophy. Comparing 2016 to how things were around 1990 when I first started to get into this field, it is transformed, and consciousness is gradually becoming a respectable scientific subject. I think the field has not done that by trying to reduce consciousness to a brain process.

What it's done is integrated consciousness with what we've been getting from the other sciences.

In particular, I see the science of consciousness as bringing together what we might call third-person data, the objective data we get from measuring systems from the outside. In the sciences of the mind, that's going to, in particular, be data about the brain, which you get from measuring the brain using various measurement methods and measuring behavior as you do with many of the paradigm methods of psychology. Those are all third-person or objective data about an organism, which have been very much at the center of the sciences of psychology and neuroscience over the years. But they don't exhaust the data about these things.

We also have first-person data, the data all of subjective experience. Right now I have first-person data about what it's like to be me sitting on this stage giving this talk. I'm seeing some people in the audience. I'm hearing my voice. I'm feeling just a little hungry. I'm looking forward to the exchange. And these are all data of subjective experience that need to be explained. And really, the problem of consciousness is how can we integrate these first-person data into our scientific picture of the world? How can we explain them in terms of the third-person data?

I think what the science of consciousness is really doing is not trying to reduce one of these to the other. Some people have said the world is all in the mind. They've tried to reduce all the third-person data to the first-person data. Others have said the reverse. "Well, consciousness is just something we can totally reduce to a process in the physical world. We'll explain the first-person data wholly in terms of the third-person data."

But I think what the science has actually been doing is bridging the two. We study the brain and we study behavior, but at the same time, we study conscious experience. We study our own conscious experience, something we can do reflectively, or we study other people's conscious experience. Of course, we can't observe someone else's conscious experience directly, but what we do is we use the method of verbal report. We ask people what their conscious experiences are. We take it as a basic principle that unless there's some specific reason to believe otherwise, to believe they're getting it wrong, then we take their reports at face value. If Antonio tells me he's feeling a pain, then he's feeling a pain. And we use that to gather data about consciousness. And our science of consciousness is all about connecting the third-person data to the first-person data.

There's a wonderful field, which has developed within neuroscience and the neuroscience of vision, but also of many other things—of emotion, of many other aspects of perception and of thought—the study of the neural correlates of consciousness, finding those brain processes which are active when you're conscious of certain information.

So when I see a color, for example, what's the process in the brain that seems to correlate most directly with my consciousness of that color? And it needs to be active for me to consciously experience the color, so when merely unconsciously I'm affected by the color, it doesn't get activated. And people are gradually narrowing that down.

Now at the moment it's mostly a field of correlations—this area of the brain is active, we get this kind of consciousness. But as we gradually make it more and more systematic and more and more universal, then we'll eventually start moving from correlation to a certain kind of explanation, which is what we want. We want a unifying theory of how you get consciousness from processes in the brain.

Now my own view is this will never be a fully reductionist theory. We'll never say that consciousness is merely a process in the brain. We'll never explain it wholly in terms of physical processes. What we really need to do is understand the fundamental laws that bridge physical processes and consciousness.

One way to get there, I think, is via the scientific method of looking at what goes on in the brain and simultaneously investigating experience with the best methods possible. I know you're going to have a Buddhist in one of the events you're organizing one of these days. I think we could do well to bring in methodologies from the East as well as from the West. They've devoted a lot of sophistication to studying consciousness from the inside. Taking those first-person data, integrating them with third-person data to ultimately be able to induce from all the data certain theoretical principles, certain theoretical bridges, which will ultimately, in my view, be the fundamental laws that connect physical processes and consciousness.

So just as physicists say they're looking for the fundamental laws of the physical world, boiling it all down to some simple principles, my own view is something like that might be the end state for the science of consciousness as well. And the kind of work that Antonio has been doing I think has been a major contribution toward this project of inducing the fundamental laws that connect the two domains.

M GLEISER: After the opening statements, Damasio and Chalmers are able to ask each other questions.

A DAMASIO: Well, it's a sort of mixture of comment and a question. Of course, I have enormous sympathy and a lot of agreement with everything that you've just said. And when one talks about reductionism, I'm always very worried about the kind of reductionism that people may think of, in which you end up saying something like, "Well, your mind is your neurons and that's all there is," which is something that was made famous by a very great and admired friend of mine, but I entirely disagree with. Because I don't think that our minds and what we are mentally, what we are as persons, is reducible to neurons.

One has to be very careful. When one does the job of science, which is fundamentally to break things apart so that we can understand a mechanism, we have to be very careful to put the mechanism back. You separate what you study, but sometimes people forget that you have to put it back, otherwise your toy is lost. And that's the danger of bad reductionism.

When you just separate, you see what the parts are, and then you get confused and you think the phenomenon is the parts. But it's not the parts. It's the parts when they work together. And there are beautiful things that are called emergences that appear once things are together, but [that] will not be there if they're not together. So I totally agree that we're not going to have a good science of mind or consciousness or many other biological phenomena if we have that narrow reductionist perspective. So full agreement there.

On the other hand, when we wonder, "Why is it that it has to feel? Why is it that there has to be experience? Why is it that it feels like anything?" there I think there can be an answer. It doesn't mean that it solves the problem all the way down. But if you take an evolutionary perspective, there is an answer. Let me explain what I think.

The fact that we feel pain or that we feel well-being or that we feel desire is not by chance. These actually occurred as the result of a long evolutionary process in which, for example, pain had a role. And what was that role? The role was to protect an organism from danger. So if you had something equivalent to pain, first as a phenomenon that might even not be perceived at all or perceived but not in the full sense, this phenomenon would guide an organism, even a very simple organism, toward approaching or not approaching something if there was a signal of danger.

So signals of danger, signals of approach, once we had nervous systems and we had the possibility of minds and the possibility of feelings, became signals that would give us either a positive valence, a pleasant response, or a negative response, of malaise or what have you. Those signals had a value and evolution would not have been possible without those signals being present.

So I would take the view that the reason why there are valences in our minds, the reason why everything that is in our minds is accompanied by qualia, is because there is an inherent necessity for feeling to occur in order for us to have operating minds.

I would also apply evolutionary thinking to why it is that we have subjective experience. I've heard people say, "What's the point of having a subjective experience?" Well, I think that subjective experiences make it matter. For example, if none of us here had any kind of leverage from feeling, and if none of us had our mental contents referred to a subject that also has feelings, why would it matter for us that we do well or bad or that we jump off the stage? I think that we have subjective experiences and I think that there are qualia because that has mattered all along.

And it didn't start as qualia, or maybe it did, but necessarily there were at first no experiences as such. But gradually they developed in that direction and now we have them in abundance. And that's why everything, if you are in a normal state, everything you perceive is in fact accompanied by that feelingness that I think is so important.

But I still entirely agree with you that this is not just about neurons. This is about neurons and it's about an organism that happens to be alive and that happens to operate in relation to the imperative of homeostasis. And therefore it's already much more complicated than the simple nervous system. And I would be strongly against any kind of reduction or summary that would take away the living organism and of course an environment in which that living organism exists in all the constraints that have led to that.

D CHALMERS: Reductionism is of course one of these very ambiguous terms that people mostly use to define views that they are opposed to.

A DAMASIO: Right.

D CHALMERS: Just take a view that is a bit more reductionist than your own and say, "Now see? I'm actually a kinder, gentler, non-reductionist!" And certainly, there's the danger of turning reductionism into a bogey person. It's just that I think one really ought to acknowledge the existence of all kinds of complicated emergent phenomena that come when you put, for example, different parts of a physical system in various relations to one another. You get the wonderful and dreadful emergent phenomenon of, say, a hurricane that comes from meteorological—

A DAMASIO: Think of sunsets, please.

D CHALMERS: Sunsets. Well, the consciousness is doing most of the work for the sunsets. What's wonderful about those is the experience. Certainly, there are wonderful emergent phenomena, but they all ultimately seem to be matters of complicated functioning of these systems. What the sunset is doing, what the organism is doing. And even at that level of emergence, there's still the question, "Why do all those emergent dynamics give you the conscious experience?"

I liked very much one thing you said toward the end about consciousness being what makes it matter. For me, this is precisely of the essence. Without consciousness, nothing would matter. Even matter would not matter. Without consciousness, there's no mattering.

Consciousness, I think, is what gives meaning and value to our lives. And people say, "What is the function of consciousness? What does it do?" I'm still not sure of the answer to that question, but I feel like it would be belittling consciousness to find something for it to do. What consciousness does is it gives meaning to everything! It gives value to everything. But that's not exactly a scientific answer to the question.

A DAMASIO: I believe it is scientific.

D CHALMERS: Maybe we could take it in that direction.

A DAMASIO: Well, we're human beings. It's very scientific to say that because we're using our knowledge and experience and our perspective to give that answer.

M GLEISER: But then you can ask, "Why us? Why *Homo sapiens*?" One of the questions is, "Is there a point along the spectrum of snails to cats to humans at which consciousness emerges?" This is, I believe, an essential question in this discussion. Are other animals conscious? And if so, where do we draw the line, if there is one?

A DAMASIO: It's a very interesting question, and of course it's a question that you have to answer with a grain of salt, because the best answer is, "At the moment, we don't know precisely." But the likelihood that we are the only species that is conscious in the sense that David and I talked about is very low. So I think that when you look at mammals, it would be a stretch of the imagination to think that any of them are not conscious in very much the same way that we are.

Now many of them will not have the richness and scope of perspective that we do. When we are conscious in the very full sense of the term, it is accompanied by a huge amount of recall from our past memories. It benefits from widening circles of knowledge that come to our minds, and we are actually made conscious, or have an experience of many of those objects and events in our minds. We do not need to have those very rich experiences all the time in order to be conscious, but we do have such experiences frequently.

Do all nonhuman animals have that too? I believe some do, at least in part. Certainly, when you consider the great apes, the likelihood of them not having some comparable experiences is very low. Of course, they do not have language to enrich the process. That's a very important limitation. And they are not likely to have the same capacity for recollection and for dealing with unique memories at the level that we do. But this is an informed guess. We cannot be entirely sure.

Then as you go down in complexity, I think we continue to find experiences that probably are very rich. It's interesting that you mentioned fish because there was an interesting discussion last year about whether or not fish had pain. And I think the general conclusion by the majority of the scientists and philosophers that were involved in the discussion was that fish do have pain and experience pain. But does it mean that they have exactly the same experience we have, to say, being caught on a hook? Not easy to tell.

I would say—and this is at the level of a hypothesis—that in order to have feelings and in order to have access to experience in the sense we're talking about here, we do need the nervous system. Prior to the emergence of a nervous system, I would put my bets on the fact that there is sentience, but not quite consciousness in the way we have discussed. In fact, there is sentience in plants too. There are very similar processes in operation in the animal and plant world.

D CHALMERS: I recommend to you the journal called *Animal Sentience*, which uses the word sentience the way you don't like. I think they mean animal consciousness, but there have been debates and articles there about fish consciousness, octopus consciousness. A lot of people have been coming out in favor of octopus consciousness, and indeed recently, insect consciousness. So be careful. Next time you eat that octopus on your plate, think about it. *[laughter]*

A DAMASIO: That's actually a very interesting point, because there are plenty of good arguments to make for invertebrates having consciousness. And, by the way, Hanna and I did write a paper on fish consciousness in the very first issue of the *Journal of Animal Sentience*.

D CHALMERS: This does raise a question, which I think is worth bringing up here, maybe to clarify your view of these matters. My colleague, Ned Block, wrote a review of your last book, *Self Comes to Mind*, six years ago, where he accused you of holding the view that consciousness always requires self-consciousness. And then the thought was, fish can have conscious experiences. To say a fish feels pain, then it's having a conscious experience in the sense of mental experience or subjective experience, even though it's probably not conscious of itself. So, would you like to reject the view that consciousness always involves self-consciousness?

Because certainly I think maybe in your past work, that's at least been a difference in emphasis between us. You've tended to emphasize the role of the self in consciousness.

M GLEISER: In summary, the question here seems to be: Does consciousness require awareness of the self?

D CHALMERS: My view is I think certainly it doesn't require anything like explicit awareness of the self. Maybe there's some background awareness of the self that potentially suffuses all consciousness, but I think a fish doesn't have any self-consciousness in any sophisticated sense, certainly.

A DAMASIO: I agree. Whatever the fish has is not, we assume, very sophisticated, and it does not require an explicit and integrated awareness of itself as an entity. But I actually have never held that view and I believe that's a misinterpretation of what I wrote in *Self Comes to Mind*, although in humans it is often the case that consciousness and explicit awareness of the self go together.

M GLEISER: Shifting gears a bit, here is a very different question on the nature of consciousness. I'd love to hear what you both have to say about it: Most discussions of consciousness seem to start with an elaborate definition of terms. To what extent does the medium of language get in the way?

To what extent does the medium of language, meaning when you start to qualify things with words, cause you to start losing precision? It happens in quantum physics. When you start to describe with words quantum uncertainty and superposition and entanglement, those terms don't mean the same thing for people as the related mathematics does.

D CHALMERS: But now you're contrasting words with numbers. You can do mathematics for quantum mechanics. But that's just another language. It's a more precise language that we've developed for physics.

M GLEISER: But again, in physics there is a severe limitation when you translate some mathematical complexity into words. Translation is a form of betrayal, even if often unavoidable and useful. I mean, popular science books would be useless if covered with mathematical formulas. The question, however, is whether we are running into the same problem with consciousness when we use language to describe, as you say, sub-jective things, which are extremely difficult to describe with words.

A DAMASIO: Yes, but we have to. What other option do we have?

M GLEISER: Well, a physicist would say math, but you can't do that with neuroscience.

D CHALMERS: Let's dance at people.

A DAMASIO: Imagine that we had come here on stage and had just been acting or drawing.

M GLEISER: No, no, this is a more serious question.

A DAMASIO: It's a serious answer.

D CHALMERS: I agree it's a serious question, but I think it's not lan-guage versus nonlanguage. It's better language. There are good languages and bad languages describing things.

M GLEISER: There you go!

D CHALMERS: I'm not saying that English is good and Portuguese is bad, but it could be we can find something that is better than both of them, as we have in physics. We've found a mathematical language, which is very good. We don't yet have a mathematics of consciousness, a vocabulary for—

M GLEISER: We don't, but let me push you a little bit because I was a lit-tle confused when you talked about needing something beyond reduc-tionism to explain what consciousness is, but we don't know what that is. I mean, in physics, we talk about collective emergence and complexity theory and stuff like that, but is that what you mean or is there some-thing more concrete?

D CHALMERS: I would say in physics, you explain some things in terms of more fundamental things. You explain a molecule in terms of atoms and so on. But in physics, we're very used to taking certain things as fundamental. Every physical theory takes some things as fundamental. Newtonian physics will take space and time and mass as fundamental. Maxwell adds charge. Quantum mechanics takes a wave function and space-time as fundamental. String theories are trying to get underneath space and time. But everyone needs to take something as fundamental and not reduce it. Once Newton took space and time as fundamental elements of his theory, and mass and so on, then he could give us equations cast in terms of those things.

I think likewise, my view is that we have to take consciousness as a fundamental element of nature. And once we've done this, then we could, for example, start to develop the equations that consciousness satisfies or the neuroscience that connects consciousness to physical processes in the brain. People now are taking up this project. Giulio Tononi has the integrated information theory of consciousness. It's highly controversial, but it has this form of trying to find mathematical equations that connect physical processes to consciousness. So I see it as analogous to what goes on in physics. It's a mathematics that takes a notion of consciousness as fundamental.

M GLEISER: So you really think that is a possibility, that you can get that through this sort of high-level mathematical approach?

D CHALMERS: I don't know whether it's got to be purely mathematical. I mean, phenomenology is extremely important, as Antonio was saying, and that's more than mathematics, but it's at least partly mathematical.

M GLEISER: OK. What do you think, Antonio?

A DAMASIO: Just going back to the language, we would have to be careful. I think that there's no other way but to try to define terms as well as possible. And if it gets boring, too bad. But if we don't do that, we can really be in great confusion. And I've seen it happen and probably most of you in this audience have seen that happen too.

So, I'm perfectly in favor of the mathematization of any kind of theoretical approach. The issue is whether or not we may lose contact with the phenomena that we're trying to explain.

There's something very real that you describe very well and I describe very well that has to do with this subjective experience and the fact that there is this feeling that accompanies say, a certain percept, and the fact that the process is related to our biological condition. We are, after all, alive. We are vulnerable. We have a beginning. We have an end. And we are the result of a long collection of adaptations over billions of years. We cannot lose sight of those phenomena. If the equations will reduce too much—because that's another form of reductionism—if they will not reduce phenomena such that we lose sight of what we are, I believe it is fine. But the risk of missing the phenomena and just getting an alternative representation is there.

Perhaps we can connect this issue with some of the current efforts in artificial intelligence, which I think are going in the wrong direction. The direction can be correct if the result is to produce amazingly intelligent artifacts that can be even much more effective and intelligent than we are. For example, if the goal is to produce a program that will win against a champion of Go or a chess champion, that's fine. AI can produce results that few of us would be capable of producing. In fact, at some point that will probably be reached very quickly, it will be able to produce results that no living person can produce.

Now, that is all well and good, provided we do not mistake that feat of artificial reasoning with a human thinking mind. From the evidence that I've seen so far, human intelligence and artificial intelligence are different. It should be studied, of course, but that's not our way. We are what we are as a result of an evolutionary process, a slow tinkering process that has taken countless generations. Only by chance will the natural and the artificial coincide in terms of structure, design, or operating capacity.

D CHALMERS: I'm curious about what you think would happen if we had an artificially intelligent simulation of those pain-signaling processes you were talking about, those processes which in humans, in your view, somehow give rise to the feeling of pain. Just say we had a very good computational process for the same kind of signaling. In your view, would that also give rise to the feeling of pain?

A DAMASIO: No.

D CHALMERS: And what's the key difference?

A DAMASIO: My prediction is that it wouldn't. I think that there would be something missing and that something has to do with—let's put it in good literary terms—the living flesh. In the same way that there's something that is like to be conscious, there's something that is like to be a living organism based on carbon and a few other fundamental elements, and to be built the way we are.

D CHALMERS: So what is it about flesh, when you get all that going in flesh but not in silicon, that now gives you the consciousness?

A DAMASIO: Well, to begin with, we get vulnerability. Our flesh is actively struggling not to be destroyed by its own entropy. It is vulnerable in a way that silicon is not, taken at the same scale. You also get a variety of very interesting molecular and cellular processes. For example, one of the things that is so interesting is that when you try to explain feelingness—leave aside consciousness for a moment—there are several new things that we are finding about the interaction of different cellular elements and different molecular elements in relation to those cells and in relation to different ages and hierarchies of the nervous system. The neurons and systems that support feelings have a different evolutionary history and different anatomical and physiological properties. Unlike cells that support cognition, for example, they lack myelin. Chemical molecules interact differently with them.

D CHALMERS: And simulating all those molecules in a process of artificial evolution is not going to do it for you?

A DAMASIO: How do you, Mr. Physicist, feel about the simulation of molecules? What is that?

M GLEISER: That's a great question. That's the big one-billion-euro grant, right? I mean, the biggest grant ever given to create a computer simulation of the brain that is incredibly complete, not just at the level of the neuronal map of what's going on, but at the level of the flow of all the neurotransmitters that flow from synapse to synapse—and there is something called the connectome which is sort of a picture of all the neurons and the synapses, sort of like a map, a subway map of the brain, with about 90 billion neurons and thousands of synapses per neuron.

I think the problem with this picture is that it assumes that we can actually get all the required information. The very nature of science is to always get incomplete information about what's going on. We never get the whole picture, right? We get part of the picture. And so getting part of the picture of a brain means not getting the full picture of the brain, which means not simulating a whole brain but simulating something else, which is what you were saying. Maybe you can get a fraction of a simulation that will do the pain stimulus.

D CHALMERS: I suspect Antonio's going to say that even if you could get all the information there about the brain at the lowest levels and you simulated that, that still wouldn't be the same as the real thing. And in particular, you wouldn't then get the experience of—

A DAMASIO: I'm not sure. I would be perfectly fine if in fact that could be produced. In fact, we have a planned set of experiments that addresses precisely that issue, and it is done with computation, of course. The question is, can you create artifacts that have any semblance of homeostatic features, because right now, the idea is that, no, you can't. And people are not that interested in doing it because, in essence, it would make the artifacts less efficient. But that simply betrays the very different approach that we must take when we are trying to develop artifacts that can be useful to us and have a particular destiny in terms of their function.

But we living things are not like that. We are now giving lots of tasks to our incredible brains that they were not designed for. We're not designed to be discussing consciousness. We ended up designed, as it were, to do such things as staying alive and healthy and having a surplus of energy at the end of the day.

2

THE NATURE OF REALITY

A Dialogue Between a Buddhist Scholar and a
Theoretical Physicist

MGLEISER: B. Alan Wallace is an expert in Tibetan Buddhism,
having been ordained as a monk by the Dalai Lama himself,
after studying with him for about 14 years. His many books
discuss Eastern, Western, scientific, philosophical, and contemplative
modes of inquiry, often focusing on relationships between the sciences
and Buddhism.

Alan is the founder of the Santa Barbara Institute for Consciousness Studies and is very active as a lecturer and instructor all over the
world. He spends only three months a year in Santa Barbara. The rest
of the time he travels, presenting workshops in places from Santa Barbara to Tuscany and beyond. He has a bachelor's degree in physics from
Amherst College, and a PhD in religious studies from Stanford. He is
definitely a person to listen to when we want to explore relationships
between the sciences and spirituality, in particular from the Tibetan
Buddhist tradition.

Sean Carroll is a world-renowned theoretical physicist, specializing
in cosmology, something called quantum field theory, and general relativity. He is currently a research professor at the California Institute of
Technology, in Pasadena. Sean has written many articles and essays for
Nature, the *New York Times*, *New Scientist*, etc. He has authored four
popular science books, the latest being *Something Deeply Hidden*, on

the interpretation of quantum mechanics, a very controversial subject. He has appeared on many TV shows about cosmology, and science in general, on the History Channel, on *Through the Worm Hole with Morgan Freeman*, and on the *Colbert Report*," which he survived to tell the tale of, not a small feat. His research focuses on the physics of the very early universe and the nature of the mysterious dark energy. His PhD is from Harvard, and he was a post-doctoral fellow at MIT and the Kavli Institute for Theoretical Physics in Santa Barbara, a few years after I was there. He was a professor at the University of Chicago, and now he is a professor at Cal Tech.

M GLEISER: This topic is very modest. The nature of reality, something very simple to deal with in a short period of time. Reality is much more than just what we perceive with our senses. We may think that what's happening right now is what we call reality. Well, that's a very small sliver of what reality is. In fact, even defining the word *reality* is quite complex. How do you know what is real? Is there such a thing as ultimate reality? And if there is, can we ever get to it? Is physics the way? The only way? Is spirituality? A combination of both? Here we have a unique opportunity to listen to a theoretical physicist and a Buddhist scholar explain their perspectives on the nature of reality and ask questions of one another.

S CARROLL: Right, the nature of reality, I want it on the record that this was not my topic choice. I didn't vet it, but it seems maybe a little presumptuous or intimidating to talk about the very nature of reality. Especially because as scientists, we instantly talk about what we don't know. I hope that no one thinks that they took a physics course, and they were doing inclined planes and pendulums rocking back and forth, and that's what professional physicists do all day. We don't just go over the stuff we know, over and over again. We go right to the stuff we don't know. We love talking about what happened at the big bang, or how to build a quantum computer.

When you talk about the nature of reality, good! There's plenty that we don't know. Part of what I want to argue for today is that there are some things that we do know. In fact, there are interesting things that we do know that we don't always tell you about. I want to make two points if that's possible, and one of them will be perfectly obvious and trivial, and you'll wonder why I'm wasting your time. The other one you will not even agree with. You will not believe what I tell you, and that's OK. You can go home, and you can think about it.

The first point is the obvious one. The universe, the nature of reality, appears to us in layers. It does not appear to us as one single unified thing that we talk to in a single way. This is a wonderful opportunity for engagement, cross-disciplinary engagement, with the humanities and the sciences. What I mean is the sort of very simple thing: the chair I'm sitting on right now, there's a way of talking about the chair. It's made of cloth and fabric and wood and metal, and that's a perfectly legitimate way of talking about the chair.

There's another way, which says the chair is a collection of atoms, of elementary particles held together by the laws of physics. And that is also a perfectly valid way of talking about the chair. The fact that there's one way of talking does not invalidate the other way of talking. They had better be compatible with each other, they [had] better be consistent. You can't say that there's one true fact that you derive from one way of thinking about the chair, one vocabulary, or one story you're telling, and an incompatible different story you're telling the other way. The feature of reality that is fascinating and worthy of our consideration is that there are many overlapping, seemingly different, but ultimately compatible ways of talking about it.

That extends not only to describing the world, but to understanding it at a deep level. We can think about the iPhone. This is my favorite example. My iPhone, what does it do? How do I talk about it? I know that if I push a certain button I can talk to Siri on my iPhone. There is a way of talking about the phone—that there are apps, and I push buttons, and I can get Google Maps, and I can go places. There's another way of talking, which is again the language of fundamental physics, that that phone is a collection of elementary particles and forces.

It turns out I understand the phone in both of those languages. I know what to do to push the buttons to get Google Maps up, and I can write down the equations to tell you what the electrons do. There's also an intermediate layer, a way of talking about the iPhone, which is that it's full of electronics, that there [are] integrated circuits, and there [are] wires and batteries, and things like that. If you ask me about that, I have no clue what is going on. How does the computational force actually work to make Google Maps figure out where to go? I have no idea what that is. It is a perfectly valid way of talking about the phone.

There [are] three layers. The layers of the fundamental particles, the layers of the electronics and the computational logic that powers the phone, and there's the layer of the use, the application in our human world. They must be compatible with each other, and it's possible that we understand some of them and don't understand others.

We all know how to get Google Maps up. My friends and I know how to write the equation for the electron. There are other people who know how to program the thing, and make it go. The fact that I don't know that layer does not mean that I'm worried that in fact when I hit the button and talk to Siri, that somehow maybe there's a little person in there named Siri. That would not be what I learned when my friends who understand the electronics told me, but what do they know? I don't understand anything about that. I'm not worried about that. Even though I don't know the details of the electronics and the programming, I'm convinced that they are perfectly compatible with the underlying atoms and particles, as well as with the emergent layer of a little phone that can give me directions to go wherever I want to go.

The universe is like that, and there are things we understand about it, and things we don't understand about it. There [are] two temptations that work in opposite directions but are very strong. There's a temptation to say "You know what? We understand everything," or "We don't understand everything today, but within a few years we'll understand everything."

This is a very common move in the history of physics especially, from the 1800s up until very recently. You could talk to a physicist, and they're like "You know, there [are] things we don't know, but it's just dotting i's and crossing t's. Any day now, we'll understand mostly everything."

Very embarrassing to say those things. They never turn out to be true. I'm not going to say that, OK? But it's tempting. There's another temptation, which is to say, "We don't understand anything." People, scientists, talk about dark matter, and dark energy, and the big bang, and quantum mechanics . . . They don't understand any of that stuff. Therefore, anything goes. Anything could be true. Who knows? That is just as easy, and lazy, and wrong, as saying that we understand everything.

The hard, courageous thing is to say, "Well, we understand some things. We don't understand other things, and here is the dividing line between them." That's difficult, but that's what we have to do if we're honest about how we understand the world.

All that's trivial and obvious, and you understood that already. Here's the surprising thing that we don't always tell you. The world appears to us in layers. There is a layer of understanding the nature of reality, which has a certain domain of validity that includes everything in this room, which we understand perfectly. It's not the layer of human interactions. If you meet a person and you're trying to figure [out] what their motivations are; are they nice? We don't understand that. We're not going to understand that any time soon.

There are many things that we don't understand, but there's a layer which is once again, that layer of the elementary particles and forces. You and I contain atoms in our bodies; that we all agree on. Atoms are made of particles, and we know what those particles are. There [are] protons, there [are] neutrons, there [are] electrons, and there [are] forces pushing them around. There's electromagnetism, gravity, there [are] nuclear forces. The dramatic claim that I want to make that rarely gets articulated as well as it should is that that layer, that language we have of talking about what's happening in this room right now, where it's particles interacting through the elementary forces of nature, we have that done. We have that figured out. We discovered the Higgs boson in 2012. We can argue whether or not we needed to discover that to count for this claim that I'm making.

But it's true, we know all of the particles that you are made of. Not only by that do I mean that we've already discovered the particles we know are in you, and me and the table and the floor and for that matter, the sun, the moon, and the stars. But we are convinced we will never, ever, ever discover new particles that are relevant to what is happening in this room, right now.

I'm saying that in very, very careful language because of course, we hope we will discover some kinds of new particles. We discovered the Higgs in 2012. We discovered the top quark in the late 1990s. Particle physicists like to discover new things. That's what they do for a living. We think that there is at least one particle that we haven't discovered yet, which makes up the dark matter, which is most of the matter in the universe.

What I'm trying to argue is that even when we discover the dark matter, it will not be relevant to understanding the physics of you and me. To understand what we are and how we behave. The secret to our biology, our life, our nature as organic beings, our thinking, our consciousness is not going to be found in changing those laws of physics. We know what the particles are; the fields, the forces that hold them together. And we know how they behave.

Of course, don't take my word for it. I'm sure you haven't, I don't need to tell you that. There are reasons why we think this. Again, it's not just that we found some particles, and we're pretty sure we're done. That would be very, very lazy, sloppy thinking if that were true. There's something very, very interesting, and powerful, about the way we think about this particular way of talking about nature. The way we think about talking about these particles and these forces. There is a whole thing . . . How long do I have to talk? You said three hours? There's a whole thing called quantum field theory, which Marcelo alluded to. A way of thinking about how these particles interact, that really puts very, very stringent constraints on what they can do.

Here's how these constraints work. Imagine there was some particle that was really important for breathing. No one believes that's true, but just to take an anodyne example. Imagine there's some new force of nature, or some new elementary particle that was really crucial to understanding how we breathe. You could say what I'm proposing is that there is some particle, and there's some number of those particles in our bodies, and it helps with our respiration.

What does that mean? That means that the particle exists, and it means that it interacts with the particles that we are made of in a noticeable way. The feature of quantum field theory is that if there exists a particle that interacts with us, then we can make it. We can create every particle that we interact with. How? By smashing other particles together.

When we found that Higgs boson in 2012, how did we find it? We smashed together protons with each other, and we created Higgs bosons. It's not that the Higgs boson was locked inside the proton. The Higgs boson is over a hundred times heavier than a proton is. But we smashed these protons together with so much force, that when we say "particles" in physics, what we really mean are vibrations in these quantum fields. The vibrations in these fields overlapped and went crazy and started all the other fields in the universe vibrating, and one of those was a little Higgs boson, which we were able to detect.

To be fair, we actually did not detect the Higgs boson, we detect the things that the Higgs boson decays into. The Higgs boson decays away in about one zeptosecond, which is a very short period of time, and we will never see a Higgs boson, it just decays away, like that. It's very hard to make. We've smashed enough particles together, in enough different ways, to say with confidence that if there are other particles out there, which surely there are, then either they decay away as soon as you make them, or they interact with us so weakly that we can't make them.

That's what we think is the case with the dark matter. It's very heavy, very hard to make, and there [are] probably dark matter particles going through this room right now, but they go right through you. They don't interact with you. They will therefore not be important for explaining how you breathe, or how you live, or how you think.

There's this one layer, which Nobel laureate Frank Wilczek has called the core theory, the theory that takes into account these elementary particles, and the forces of nature, including gravity, and electromagnetism and so forth, and my claim is that we understand that one layer.

That doesn't mean we understand the bottom layer of reality. There could be layers underneath. I'm pretty sure there are. I'm pretty sure we don't understand the nature of space and time themselves. That's OK. We don't understand the layers on top of the elementary particles. We don't understand chemistry, biology, sociology, politics, my goodness, we don't understand very well, at all. That's also OK. We don't even understand the layer sideways from the elementary particles that we know and love. We don't understand the dark matter and other astro-physical phenomena.

It's actually a very grandiose, bombastic-sounding claim, that when you really think about it carefully, is quite restricted in what we're saying. We understand certain elementary particles, in a certain regime of what they can possibly do. It's just that that regime includes everything that you and I ever do, and everything that we see around us in our everyday lives.

The way that I say it is the laws of physics underlying our everyday lives are completely known. Again, that doesn't mean we understand all the emergent layers on top. In fact, there's a lot to learn, when you talk about elementary particles ganging up to make complex, emergent phenomena, macroscopically, like you and me.

If you meet someone, if you have a date with someone, you go on a blind date. You [have] never met this person. You're talking to them, and they say "Well, tell me about yourself." You would not start listing all of your atoms, and say "Well, this atom does this, and this atom does that." If you were an undergraduate at Cal Tech, that might be kind of sexy. But for most of us, we tell a story that is at a much higher level than that. We don't know the ultimate theory of reality at that those higher levels. In exactly the same way that I don't know how to talk about the integrated circuit and the programming in my iPhone. But they're both based on these particles that we've made as physicists and what they do.

There are really important implications of this fact, that the laws of physics underlying our everyday lives are completely known. One is that you cannot bend the spoons with your mind. You cannot hold up . . . You may have seen people claim to do this. Hold up a spoon, just look at it, and it bends just by the force of you thinking about it. You can't do that, sorry. I can't name names, because litigation always follows. The idea [is] there is a force that I can focus with my brain that somehow bends the spoon; well look, I know what all the forces could be. I know what all the particles are that make up my brain, and all the particles are that make up the spoon. I know there's none that can reach over there and bend the spoon. Most people are happy with that.

It also means that there's no life after death. That's where people sort of get off the bus, and then stop listening to me, but I think it's true. This is my belief, and I'm giving you the arguments why I believe it. Because we know what you are. We know the stuff that makes you up. We don't know how that stuff comes together, to make you who you are in the personal sense. But we know that the stuff that makes you up, we know all of it, and we know that when you die, it's still there. It doesn't go anywhere. There's no way for the information that makes you, You, the memories that you have, the personality traits, all those charming little quirks that make you, You. There's no way for that to go anywhere, because we know what you're made of.

So some people take this fact about what we understand about the world as disenchanting, or it's not romantic, or it makes them sad, because they think that something is lacking. It seems very clinical and dry, and only a physicist could love this world, where we're all made of particles, obeying the laws of physics.

I can even write down the equation. In fact, you can buy t-shirts with the equation on it, that governs all the particles and fields inside you. Is it somehow less of a lovable universe when you live in a universe for which that part of reality is described in that way? I don't think so. I think that the ramifications of this way of thinking are enormous, and if anything, we don't take them seriously enough. But I don't think that it's dispiriting in any way.

The reason why I like to dwell on the life after death thing is because that's where it really gets us the closest in our hearts. That's what we have trouble accepting, very often. You can go through the math. It's a fun thing to do. It turns out that all different kinds of mammals here on Earth live for the same number of heartbeats. If you're a more massive animal, if you're like an elephant or a whale, you live longer, but your heart beats more slowly. If you're a tiny little mouse, or a rodent, you live for shorter, but your heart beats faster. It exactly cancels. All the mammals on Earth live for about one and a half billion heartbeats.

We human beings, if you do the math for your heart rate, that works out to about 35 or 40 years, which is more or less how long we lived in the state of nature. But now, we have Obamacare for a little while. Now, we live for seventy or eighty years. The average human being gets three billion heartbeats, and if you believe the story I'm telling you about what you are made of, and the fact that we know what happens to that stuff when you die, then those are all you get, three billion heartbeats.

Of course, I'm not saying anything different than saying the average human being lives for seventy or eighty years, but putting it in terms of heartbeats I think evokes a certain feeling, because years take a long time to go by, and heartbeats . . . Look, you've squandered several thousand heartbeats just since you sat here, OK? Three billion is a big number, but it's not that big. To think that we only have three billion heartbeats, and we're squandering them roughly once a second, and that's all we're going to get, I don't find that saddening or dispiriting at all. I think it's a challenge, but it's liberating. It really focuses the mind. It's not a dress rehearsal, this life that we're living here. We have a certain amount of it, and every heartbeat counts.

The electrons and the particles that I'm made of, they're going to follow their equations. That's what they're going to do. But the fact that I, as a macroscopic person, have feelings, and thoughts, and choices, that doesn't go away, anymore than the fact that I can talk to Siri goes away because I know that the phone is made of those particles. I have the ability to make choices. I have the ability to make something of my life. We human beings have deployed that ability to understand this universe that we're in. To understand a lot about elementary particles, a tiny bit about chemistry, almost nothing about politics. What that means is there's plenty of room to do more things. We're nowhere near understanding all of the nature of reality, but we understand a little bit. That's an amazing intellectual accomplishment, and I'm very fascinated to see what the things are that we're going to learn next.

B. ALAN WALLACE: In Sean Carroll's very provocative book, *The Big Picture*, which I have read very carefully, taking twenty-six pages of single-spaced notes, I feel as if I've seen into your mind, speaking only metaphorically, of course.

You quote the poet Muriel Rukeyser with a delightful phrase, "The universe is made of stories, not of atoms," from her marvelous poem, "The Speed of Darkness." Sean himself adds here, "The world is what exists and what happens, but we gain enormous insight by talking about it—telling its story—in different ways." We're on common ground so far.

This is a good place to start, with the acknowledgment that there are multiple stories to be told, and the willingness to listen to diverse ones that may be complimentary and that may challenge our assumptions. I think this is inherently a worthwhile thing to do.

In the Buddhist view, the defining characteristics of human beings— what makes us special or distinctive—are that we are able to understand meaningful information, and we're able to impart it as well. In short, we are storytellers. We tell many stories. We've heard a very good one here from Sean, based on a lot of science, but it is one of many stories. I would suggest that all the stories we tell ourselves, that we tell each other about the physical universe, life, the mind and so forth, begin with certain assumptions that we believe to be true, but that we cannot prove with empirical evidence or logic.

We start with assumptions, beliefs, or axioms. On this basis, we develop theories. And then, throughout the course of our lives, we choose what we attend to, where we focus our attention. If you're a professional physicist, you attend to physical things. And our world as we envision it corresponds directly to what we attend to, and it excludes what we ignore. As the great pioneer of modern psychology William James wrote, "The subjects adhered to become real subjects, the attributes adhered to real attributes, the existence adhered to real existence; whilst the subjects disregarded become imaginary subjects, the attributes disregarded erroneous attributes, and the existence disregarded an existence in no man's land, in the limbo 'where footless fancies dwell.' . . . Habitually and practically, we do not *count* these disregarded things as existents at all. They are not even treated as appearances; they are treated as if they were mere waste, equivalent to nothing at all."[1] In short, "Our belief and attention are the same fact. For the moment, what we attend to is reality."[2]

In his book and during his lecture this evening, Sean alludes many times to the underlying nature of reality. This is a critical point: What underlies what? What are our starting assumptions? What are we attending to? What kinds of questions are we posing to nature, and how do we seek to get the answers to them? In his book, Sean makes his own beliefs very clear: "there is only one world, the natural world, exhibiting patterns we call the 'laws of nature,' and which is discoverable by the methods of science and empirical investigation. There is no separate realm of the supernatural, spiritual, or divine; nor is there any cosmic teleology or transcendent purpose inherent in the nature of the universe or in human life. 'Life' and 'consciousness' do not denote essences distinct from matter; they are ways of talking about phenomena that emerge from the interplay of extraordinarily complex systems."[3]

This is the kind of assertion one would expect from a physicist. This is one of the many stories we tell ourselves about the nature of reality. But it's not the only story, not even the only scientific story. Our notions of what underlies what are based on our starting assumptions, the questions we pose, the focus of our attention, and the conceptual framework in which we make sense of our observations.

Sean's book reveals three unquestioned assumptions: determinism (the future follows uniquely from the present), realism (there is an objective real world, independent of any observer), and physicalism (the world is purely physical). These are beliefs one might expect from a physicist, but probably not from a poet, a mathematician, or a contemplative, for example.

While many people embrace these metaphysical beliefs and conflate them with the empirical facts of science, it's important that we take note of other worldviews advocated by leading scientists, mathematicians, and philosophers [who] do not adhere to those beliefs.

The eminent mathematician Roger Penrose, for instance, points out in his book *The Emperor's New Mind* that the fact that the laws of physics are precisely mathematical suggests that the physical universe may emerge from an underlying dimension that is purely mathematical. This view traces back to Plato and the Pythagoreans. It raises the question of why the laws of nature should be mathematical at all, let alone precisely mathematical. Philosophers, mathematicians, and scientists tracing back twenty-five hundred years to Pythagoras have posited that maybe the underlying reality is not physical at all. It is not composed of particles or forces but is a dimension of reality that consists of numbers and pure geometrical forms.

Roger Penrose rejects Sean's belief that mathematicians start out with axioms, and then merely make logical deductions about possible realities. According to Penrose, mathematicians are actually exploring a dimension of reality that's pure math. They make discoveries, and they inter-subjectively validate each other's discoveries of a deeper, underlying realm of reality that is mathematical in nature.

Werner Heisenberg, one of the great pioneers of quantum mechanics, wrote, "With regard to this question, modern physics takes a definite stand *against* the materialism of Democritus and *for* Plato and the Pythagoreans."[4] This is a story that presents mathematics, not physics, as the most fundamental of sciences. It is quite different from the materialist vision promoted by Sean, but it is no less scientific.

In addition to the precisely mathematical nature of the laws of the universe, physicists have noticed that the fundamental physical constants of nature are such that had they been even slightly different, the universe would have been void of intelligent life. This has led some physicists such as John D. Barrow and Frank J. Tipler to conclude, "There exists one possible Universe 'designed' with the goal of generating and sustaining 'observers.'"[5] And the eminent theoretical physicist John A. Wheeler adds, "According to the principle, a life-giving factor lies at the centre of the whole machinery and design of the world."[6]

In other words, this universe is life friendly. Underlying the inanimate physical universe there may be a primordial life force that determines that the cosmos evolves in such a way that it can support the emergence of living beings. A theist might attribute this intelligent design to God, who created this universe in order to bring forth intelligent life. In this case, biology, not physics, would be the fundamental science.

There are other great physicists, including Max Planck, the founder of quantum mechanics, who conclude that mind is the fundamental reality underlying the physical world. He wrote, "All matter originates and exists only by virtue of a force which brings the particle of an atom to vibration and holds this most minute solar system of the atom together. We must assume behind this force the existence of a conscious and intelligent mind. This mind is the matrix of all matter."[7] Einstein expressed a similar view when he declared that there is a "superior mind that reveals itself in the world of experience."[8] To cite one more example, Sir James Jeans, one of the great pioneers of cosmology in Great Britain, wrote, "the Universe begins to look more like a great thought than like a great machine. Mind no longer appears to be an accidental intruder into the realm of matter . . . we ought rather hail it as the creator and governor of the realm of matter."[9] If a superior mind, such as the mind of God as envisioned by Spinoza and advocated by Einstein, is indeed the underlying reality of the universe, then theology would be the most fundamental of sciences, as was widely believed in the medieval era.

There are still more alternative views regarding the underlying nature of reality that have been proposed by prominent physicists. One of them is Anton Zeilinger, a leading experimental physicist at the cutting edge of exploring the nature of quantum reality. Following the principles of quantum cosmology originally proposed by John Wheeler, Zeilinger writes, ""One may be tempted to assume that whenever we ask questions of nature, of the world there outside, there is reality existing independently of what can be said about it. We will now claim that such a position is void of any meaning. It is obvious that any property or feature of reality 'out there' can only be based on information we receive. There cannot be any statement whatsoever about the world or about reality that is not based on such information. It therefore follows that the concept of a reality without at least the ability in principle to make statements about it to obtain information about its features is devoid of any possibility of confirmation or proof. This implies that the distinction between information, that is knowledge, and reality is devoid of any meaning. Evidently what we are talking about is again a unification of very different concepts. The reader might recall that unification is one of the main themes of the development of modern science."[10]

According to this view, meaningful information, not matter or energy, is fundamental to universe. This is the theory that bits of information are primary, while all other conceptual constructs, such as space, time, particles, fields, and forces, are derivative. This is summarized in John Wheeler's phrase "it from bit." In his own words, "It from bit symbolizes the idea that every item of the physical world has at bottom—a very deep bottom, in most instances—an immaterial source and explanation; that which we call reality arises in the last analysis from the posing of yes-no questions and the registering of equipment-evoked responses; in short, that all things physical are information-theoretic in origin and that this is a *participatory universe.*"[11]

This implies that information science is the most fundamental of all sciences, and it explicitly raises the question as to the role of the observer and the conscious "information processor" in nature.

To highlight yet another view regarding the underlying nature of reality, Stanford physicist Andrei Linde suggests that consciousness may be fundamental to the world of nature, rather than an emergent property of complex configurations of atoms. He suggests another possibility of what's underlying. He writes, "The current scientific model of the material world obeying laws of physics has been so successful that we forget about our starting point—as conscious observers—and conclude that matter is the only reality and that perceptions are only helpful for describing it. But, in fact, we are substituting the *reality* of our experience of the universe with a conceptually contrived *belief* in an independently existing material world."[12] He then raises the question, "Is it possible that consciousness, like space-time, has its own intrinsic degrees of freedom, and that neglecting these will lead to a description of the universe that is fundamentally incomplete? What if our perceptions are as real (or maybe, in a certain sense, are even more real) than material objects?"[13]

The core theory, as presented by Sean Carroll, may be fundamentally incomplete after all, for it ignores something that is absolutely essential: the consciousness with which we observe and understand the natural world. Its nature and origins and its role in the universe remain unknown and largely overlooked in modern science as a whole. If consciousness underlies the world of nature, then the cognitive sciences should be the most fundamental of sciences, but they have left us in the dark regarding the nature and origins of consciousness.

Finally, there are distinguished physicists who believe that the underlying nature of reality is simply unknown. Freeman J. Dyson expressed this view when he wrote, "I am reluctant to engage in discussions about the meaning of quantum theory, because I find that the experts in this area have a tendency to speak with dogmatic certainty, each of them convinced that one particular solution to the problem has a unique claim to be the final truth . . . As a physicist, I am much more impressed by our ignorance than by our knowledge."[14]

In a similar vein, Sean Carroll writes: "Quantum mechanics is the deepest and most fundamental picture of the world we now have, but what it says about reality is utterly uncertain." Regardless of the uncertainty of whatever quantum mechanics has to say about the nature of reality as it exists independently of all measurements, Sean maintains his unquestioning belief that [it] is [the] deepest and most fundamental picture of the world we now have. This is an understandable predilection by someone who has devoted his professional career to the study of quantum mechanics. Marcelo Gleiser likewise acknowledges, "we are essentially blind to what exists at the very core of physical reality. All we have is our measurements, and they give an incomplete picture of what's really going on."[15]

Now we turn to Buddhism, segueing from Marcelo's comment that we are essentially blind to what exists at the very core of physical reality. I'd like to begin by alluding to the parable of the blind men and the elephant taught by the Buddha. In this story, there was a king who gathered together a group of blind men and presented them an elephant.

These blind men individually touched the head, ear, tusk, trunk, the side of the body, a foot, the back, the tail, and the tuft of the tail and consequently described the elephant as pot, a winnowing basket, a ploughshare, a plough, a grainery, a pillar, a mortar, a pestle, and a brush. Then, the Buddha narrated, "they began to quarrel, shouting, 'Yes, it is!' 'No, it is not!' 'An elephant is not that!' 'Yes, it's like that!' and so on, till they came to blows over the matter." Each blind man, based on his partial observation of the elephant, came up with a different, incompatible description of the elephant as a whole. No one description was wrong, but they were all incomplete and appeared to be incompatible. The Buddha concluded this parable with the observation, "Just so are these teachers and scholars holding various views blind and unseeing. . . . In their ignorance, they are by nature quarrelsome, wrangling, and disputatious, each maintaining reality is thus and thus."[16] Now it's worth noting that virtually all of the individuals I cited previously are twentieth-century, white, male physicists, cosmologists, and mathematicians. This is a very narrow bandwidth of humanity as a whole, and despite the commonality of their professional training, their views about the fundamental nature of reality are very diverse.

A core question is whether modern science is incompatible with beliefs in the supernatural, the continuity of consciousness after death, and paranormal abilities, which are widely held in diverse cultures throughout human history. Even among physicists, there is a good deal of disagreement about such issues.

There's a lot of emphasis nowadays in the West on developing "secular Buddhism." I'd like to suggest that the first secular Buddhist was the Buddha. To support this hypothesis, I shall cite a well-known discourse the Buddha gave called the *Kālāma Sutta*. The context for this discourse was a visit by the Buddha to a village whose inhabitants had heard a lot of people propounding the truth, each one with a story incompatible with the others, and each one saying, "My view alone is right." As a result, the villagers had become very skeptical toward anyone who claimed to have answers to the big questions of existence.

Rather than arguing that his views were superior to everyone else's, the Buddha surprised them by responding, "It is proper for you . . . to doubt, to be uncertain, for there are good grounds for your uncertainty. . . . Do not adopt views simply because they are the status quo, or because they've long been assumed to be true. Do not accept them based on mere rumor or because they are written in some scripture. Do not accept them based on pure conjecture, some unquestioned assumption, inconclusive reasoning, your own personal bias or others' eloquence, or because it's the view of your teacher."[17]

I believe this is an expression of healthy skepticism, and then on a positive note, he counseled the villagers to test all truth claims with their own experience and determine for themselves whether they were true or false, beneficial or harmful. He was truly an empiricist. Let's recall the various stories I briefly alluded to previously, each one advocated by white, male, twentieth-century, mainstream physicists. I find a common element to all these views that entails two kinds of blindness in modern science at large.

The first is a blindness to the possibility of profound insights into the fundamental nature of reality from any of the great civilizations of Asia over the past five thousand years. In all those stories, where do we ever hear of any references to discoveries made in the traditional cultures of China, Japan, India, southeast Asia, Tibet, or central Asia? Did those Asian civilizations—let alone indigenous cultures all over North and South America, Africa, Australia—not discover anything of significance about the nature of reality?

What I'm getting at is that this is a very narrow sliver of humanity I've cited in the previous views on the underlying nature of reality seems to regard itself as the center of the universe. Nowadays we frequently here the question raised, "Is there intelligent life elsewhere in the universe?" My answer is yeah—Asia! With our ethnocentric biases, we in the West have been overlooking discoveries made in the East over the entire course of recorded human history. That's a massive oversight. If we were still living in the nineteenth century, such an oversight would be understandable. After all, Eurocentric nations were busy conquering the rest of the world and dominating all other cultures with our own, with the sense that the West alone was truly civilized. But now, in the twenty-first century, it is inexcusable to ignore the wisdom of whole civilizations.

That's one blind spot. The second blind spot that I would like to highlight is the mind. Virtually all the diverse branches of modern science are characterized by one common characteristic, which has been instrumental in leading to their great success. With one notable exception, each one has developed sophisticated, rigorous, precise modes of measurement, of observation, of the phenomena scientists have been seeking to understand. The one exception is the mind and the entire range of mental processes and states of consciousness.

Behaviorists study the mind indirectly by studying the behavioral expressions of mental processes. Brain scientists study the mind indirectly by exploring the brain correlates of mental processes. But nowhere among the mainstream cognitive sciences is there any rigorous training in directly observing the only minds we have any direct access to—our own. We study physical processes related to mind, but scientists have developed no sophisticated, precise, replicable methods for observing, exploring, and experimenting on the mind from a first-person perspective.

The refinement of attention, mindfulness, and introspection has been a great strength of multiple contemplative traditions throughout Asia for five thousand years and in the Buddhist tradition for the past twenty-six hundred. The development of attention skills is crucial in this regard, for the mind can't be directly observed with any instrument of technology. It can be observed only introspectively.

But not only has modern Eurocentric civilization failed to develop effective methods for developing sustained, voluntary attention, we are actually getting worse, as evidenced by the increase in attention deficit/hyperactivity disorders (ADHD). As a civilization, we're slipping into a trough of ADHD from which we may never extricate ourselves. While we lack effective means for overcoming such attentional imbalances and have failed to develop methods for cultivating exceptional degrees of mental focus, this is one of the great strengths of Asian contemplative traditions, especially Buddhism, where such skills have long been used to observe all manner of mental processes and to explore multiple dimensions of consciousness.

Scientifically, we're not closely attending to the mind, so to some people, including Sean Carroll, it seems to be just a way of talking about the brain. It's not, and I can demonstrate that. It's going to take ten seconds. For ten seconds, I'm going to ask you to do something very simple. Mentally, be totally silent, don't move, don't behave, and just be aware of what's coming to mind . . .

I would say with tremendous confidence that what you just experienced was not a way of talking about the brain, because you weren't talking. It's not a way of behaving, because you're not behaving. Your experience of your own mind and other appearances to consciousness are present before talking and before deliberate, conscious behaving. The scientific community has been overlooking something—our own first-person, direct experience of our own minds—and it's been pretending as if the mind-body problem is already solved, that the mind is nothing more than what the brain does. That's just a belief, and one that has never been subjected to critical, scientific examination.

The measurement problem in quantum mechanics has not been solved since it first arose around ninety years ago. In fact, there's been no progress in this regard. Likewise, there's been no progress in solving the mind-body problem since the origin of the mind sciences in the late nineteenth century. Over the past century, we've been overlooking something, and it is a major weakness of modern science: we've overlooked the nature of the observer. We're not taking a truly scientific, open-minded, empirical approach to the nature of mind and consciousness. We've left out something in our understanding of the natural world, and that's us. Oddly enough, we are more ignorant about consciousness than we are about distant galaxies that formed ten billion years ago, more ignorant about consciousness than we are about the inner nucleus of an atom. Why? Because we're not looking.

What Buddhism brings to this historical situation is not dogma, but rather methods of rigorous, refined, first-person inquiry into mind as it is subjectively experienced, to compliment the magnificent accomplishments of objective science, which is the strength of the West.

M GLEISER: You are free to respond, Sean, then Alan.

S CARROLL: Too many things to say, obviously. We could have hours and hours of discussion. Rather than asking a question. . . . All right, let me ask one quick question. What is the difference between mind and individual minds? When you were talking about mind and the centrality of mind, do we have in mind something singular, and monistic, and cosmic, or are we thinking about that each one of us has one of these minds?

B. ALAN WALLACE: According to the story proposed by Max Planck, Einstein, and Sir James Jeans, *mind* refers to a cosmic mind. But that's not what I'm referring to in my comments regarding Buddhism. Buddhism doesn't start with a leap of faith in the existence of God, or Nirvana, or reincarnation, or karma, or Buddha. We start with our most indubitable knowledge, namely that there is consciousness and appearances to consciousness.

S CARROLL: Good, excellent. Let me make two quick points, then I think Alan can respond at any length. The two quick points—so I disagree that consciousness and the mind are central, but that's great. I mean, that's why we're here. We knew ahead of time this would happen.

And you can agree or disagree with one of us, or both of us, and that's also fine. We're trying to give each other things to think about. We're not trying to actually . . . I think this is not the last time we get to talk about the nature of reality in human history.

B. ALAN WALLACE: I expect you're right.

S CARROLL: We'll go on from there. Let me just put forward two ideas that help you understand why I don't think that mind is separate, in addition to your brain, and the particles and forces that make up your brain.

One is that it seems to me to be like the last gasp of anti-Copernicanism. Copernicus many years ago made a lot of money by saying "Maybe the Earth is not the center of the solar system. Maybe we're displaced."

Traditionally, not always, but in traditional, ancient cosmologies, the Earth had a privileged role. It was either at the center or was doing something important. Copernicus said, "Maybe it's off to the side." Kepler said, "Maybe they're ellipses." Galileo actually showed that there were things orbiting around Jupiter, not just around the Earth.

Ever since then in the history of science, there have been a number of episodes in which our importance as people to the workings of the grander cosmos has seemed to diminish. Even Copernicus thought that the sun was special, right? But now we know there [are] lots of stars out there.

It wasn't until the 1920s that we knew that there were other galaxies in the universe. Darwin showed that the existence of life, and the origin of us as human beings, is not special. It's continuous with the physical processes here on Earth.

To me, when I look at one of these images that you can see online, or get big posters of what is called the Hubble Ultra Deep Field—this is an image of galaxies, this is the picture you would get if you went out on a clear night with your camera, you point it at a blank spot [in] the sky, you click the shutter, and if your camera is attached to the Hubble Space Telescope, you will see that even the blankest parts of the skies are alive with galaxies. There are two trillion galaxies in our observable universe, [as] our best current estimate. On average, 100 billion stars per galaxy. Who knows what's beyond the observable universe? Perhaps a lot more, and to look at all that stuff—and this is a very cheap, tawdry argument;

I'm sorry for even giving it but it is very visceral—to look at all that and go "Oh yes! Consciousness must be central to this," just seems very backwards to me.

Consciousness is the tiniest sliver of an imperceptible little speck on one little thing, in a big, big collection of things, so that's hard for me to buy. That's point one.

Point two, which will be much faster—there's a chapter in my book, my favorite chapter, about Princess Elizabeth of Bohemia. She is a person who would be like a world-famous philosopher or scientist, but in the 1600s if you're female, you don't get to be a world-famous philosopher. But she did engage in a conversation with Rene Descartes, and Descartes was of course trying to articulate this idea that mind was somehow separate from the body.

Elizabeth engaged him in correspondence back and forth. Frankly, she got the better of him. He was a little snotty at first. He didn't think that a princess was really up to his level. But she kind of schooled him, pretty badly. Her basic argument was "How in the world can you explain how this immaterial, disembodied, non-located thing you're calling a mind ever affects my body, ever pushes it around? Where does that happen? Show me where it actually happens." Because we know we have bodies, whether or not the mind is separate from them. We know we have bodies.

The modern version of this is we know we have particles in us. We know we have atoms, and we know what they do. We know how they behave. We know an equation that says if you put them in a certain situation, here's what's going to happen next. There's no room in that equation for consciousness, or the mind to come and go "Oh well, if this electron is in a mind, then it moves that way."

It's completely possible that there should be. As scientists, we're very open. I could totally be wrong about everything I told you. If new evidence or argumentation comes that shows us that there's a better way of understanding the world in which that's the case, then we will change our minds, whether they're atoms or something else.

The point is it's really hard to imagine, given what we know about the stuff that is in our brains, that there is anything else pushing it around.

Does that stuff conserve energy? Is it predictable? Is it deterministic? Does it obey Schrodinger's equation? Does it have a location in space? There might be answers to all these questions, but there's also an easier answer, which is that there isn't any such stuff. The mind is hard to understand, because it's the most complicated thing we know. Of course, it's hard to understand. That doesn't seem to me to mean that we need to invoke new stuff, outside the laws of physics.

B. ALAN WALLACE: Consciousness is not *new stuff*. Whatever it is, it's not new. It's been here long before scientists started thinking about it.

Consciousness is a blind spot in the scientific view. It's as if the mind doesn't matter, because it's not matter, or it matters only if it is matter. There is in fact no evidence at all that consciousness is matter. It has no physical qualities, and it can't be measured physically. So the evidence points to the immateriality of consciousness. Nor is there any evidence that mental processes are equivalent to the correlated brain states, nor is the mind nothing more than a way of talking about the brain. It's still called the mind-body problem because it hasn't been solved.

Consciousness is not just a way of talking. Love for a spouse, for a friend, for a teacher, for humanity, is not a way of talking. There is no scientific definition of consciousness; there's no way of measuring it objectively, in anything. Scientists don't know the necessary and sufficient causes and conditions that give rise to consciousness. Are plants consciousness? Are elementary particles consciousness? When does a human fetus become conscious? The answer is we don't know, scientifically.

Since scientists don't know what are the necessary conditions that give rise to consciousness, it is mere hubris to say that, "Well, we don't know where it comes from. But we certainly [know] it terminates at death." No, you don't.

In the preface of his book *The Discovers*, the historian Daniel Boorstein wrote that the greatest obstacle to scientific discovery is not ignorance, but the illusion of knowledge. The scientific community is impoverished in terms of its understanding of consciousness, but it is rich in terms of its illusions of knowledge. The unquestioned assumption is that if there is a sufficiently complex array of neurons, then consciousness emerges from their interactions. But no one has a clue to what sufficiently complex means. It's a pseudo-hypothesis that has never been subjected to the test of empirical research. It's simply accepted as an article of faith, misleadingly presented as scientific knowledge.

Thomas Huxley, the nineteenth-century biologist who founded the Church of Scientific Materialism, to which I see Sean is a devout member, wrote, "how it is that anything so remarkable as a state of consciousness comes about as a result of irritating nervous tissue, is just as unaccountable as the appearance of the Djinn when Aladdin rubbed his lamp."[18] The notion that consciousness emerges from the chemicals and electricity in the brain is an expression of neuro-mythology. It's a mystical view of the brain that believes it is solely responsible for the emergence of all manner of subjective experience, none of which display any physical characteristics whatsoever. It's magical thinking, really.

Love is not a way of talking about neurons. We understand love by loving, not by knowing about neurons. One is not reducible to another. Only a robot (or a person emulating a robot) might propose that one is reducible to the other, because a robot doesn't have any subjective experience and doesn't really *know* anything at all.

The belief that since we can't explain the subjectively experienced mind in terms of physics, it must be nothing other than the brain, is a way of avoiding the problem, rather than facing it directly. Sean advocates determinism, the belief that the future follows uniquely from the present, and that the present is fully determined by the configuration of physical phenomena in the past. This implies that every time you feel you're making a choice, and that once you've made a decision and acted accordingly, [that] you could have acted differently is an illusion. According to this form of determinism, namely, Laplacian Determinism, which most physicists believe has been utterly repudiated by quantum physics, you could not have acted differently, because your behavior was wholly determined by the configuration of physical phenomena, and the laws of nature.

If you feel comfortable with that, I would suggest you go back to the novelist Isaac Bashevis Singer, who said when he was asked, "Do you believe in free will?" He said "Yeah, of course. Do I have any choice?" We have to be taking first-person experience far more seriously, and we haven't, because for about the last century, since John Watson and the rise of behaviorism—which corresponded to the snuffing out of the radical empiricism of William James—scientists have been focusing their attention solely on physical phenomena, namely, on behavior and the brain. But they're neglecting the hallmark of all the other successful branches of science: to directly, rigorously observe the natural phenomena you are seeking to understand. As long as cognitive scientists continue to focus only on the physical expressions and correlates of mental processes, they will remain stuck in their materialistic paradigm, and the measurement problem and the mind-body problem will remain unsolved.

I'm not here to promote some Buddhist worldview or a set of beliefs. I'm not here to promote the notion mind is separate from matter. But I am saying that our scientific reductionist, materialist way of addressing the mind now is not scientific. Not in the way that geology, particle physics, astronomy, and so forth [are], where you rigorously observe the phenomenon you're trying to see. Introspection plays almost no role, and refined, developed introspection plays no role whatsoever in the modern scientific study of the mind.

We're missing something here, and that's because the dogma of physicalism, which insists, "If you skip the first-person perspective, skip introspection, what have you missed after all?" The Buddhist view is not like the pre-Copernican belief that Earth is in the center of the universe. It's rather the view that there are countless conscious beings throughout the universe, not just those on planet Earth.

The Buddhist view is not the Abrahamic. In the Buddhist view, planet Earth is not the center of the universe. The Buddhist view is that our world is just one in a billionfold cluster of inhabited world systems, and there are countless such billionfold clusters through the universe. To relate this assertion to modern cosmology, if the average galaxy has 100 billion stars, and around one out of every hundred stars has a planet orbiting that sustains sentient life, then each galaxy would correspond to a Buddhist "billionfold cluster of inhabited worlds." We have to get over the notion that *religion* is confined to the Abrahamic religions. I don't even think of Buddhism as a religion. Rather, it's a radically empirical way to investigate the nature of reality from the inside out, of examining the mind from the within, and to explore its role in the natural world as a whole.

I'm not here to argue mind-body dualism or any other dualisms. That's just a way of talking. But I'm suggesting that we adopt a radically empirical, truly scientific approach to observing the mind, and not just talking it out of existence by saying it's just a way of talking about neurons.

M GLEISER: If there's something we can learn from tonight's conversation, it's that being human is both exhilarating and perplexing. We, at least hopefully, really enjoy the ability that we have to ask fundamental questions about existence. I would slightly disagree with Sean in being anti-Copernican—in the sense of we are just the norm in the universe, by saying that there is something that makes us humans special, because we are here and are capable of asking questions about the nature of reality and the puzzles of existence.

S CARROLL: You're a human, but I'm a robot, because I think it all comes down to neurons.

M GLEISER: Well, OK. I didn't know they could do this machine learning so well, but I guess they can since you fooled me. That's awesome.

In that sense, I think that it's a privilege to be able to be asking these questions, and to be able to reflect on all this. This issue of the binding problem, that is, of how something immaterial interacts with something material, is extremely complex. What Sean basically is saying is that we have zero evidence that there is anything there; meaning there are no mysterious forces.

You could make similar arguments about the issue of the natural versus the supernatural. Let's say you tell me that you've seen a ghost. I could reason that "Well, you've seen a ghost. This means that the ghost is an electromagnetic manifestation, so it's a perfectly real thing that you can understand through normal physics." In other words, when the immaterial interacts with the material, what is supposedly supernatural interacts with the natural and becomes part of nature. It should then be understood within that context.

The fact is that we don't have and can't have all the tools. That is the essence of science, what we don't know. I think we all agree that we really don't understand consciousness. Anybody [who] says that he or she does shouldn't be taken seriously. We had our first public dialogue about this very question with Antonio Damasio and David Chalmers. There is a group of philosophers, which are called the Mysterians—I love that name—which basically stresses this point, which is called the hard problem of consciousness. They claim that we are not equipped to understand the nature of consciousness, because we are so much within our own narrative, locked within our minds and bodies, that to understand consciousness as an objective object of study, like an electron is, is just not possible. This blind spot makes the whole idea of studying consciousness a real challenge for our current scientific narrative.

Not to say, perhaps, that consciousness is terminally impossible for us to understand. Beware of absolute statements about ultimate impossibilities. Maybe a very smart robot will come from the Bay Area, who knows, right? Or some very sophisticated alien will be able to do that one day. Or science will change in radical ways in the future. But right now, we are very confused by the whole idea.

The nature of reality is also a very confusing topic, because as Sean was saying from the beginning, there are all these different layers, and complementary ways of knowing. I think if anything, tonight, what we've learned is that we have to look at all these different points of view with a huge dose of humility, because clearly we don't understand all there is to understand. We don't even know the questions we should be asking to illuminate the path forward with any certainty.

An essential part of the pursuit of knowledge, the starting point, really, is to accept our ignorance. That's exactly what makes us want to ask questions, because we don't know. As Tom Stoppard once wrote, "It's wanting to know that makes us matter." We should celebrate not knowing, and stay curious about what we can understand about who we are.

And now we can have some questions.

QUESTION ONE: This is a question about the heartbeat that you brought up, Sean. We know that the heartbeat starts at the sinoatrial node, right? It's this electrical impulse, that starts the heart. So the heart pumps out oxygen to the body, including the brain. I asked an electrophysiologist "How does the heart know to cause an electrical impulse to start the heart going?" She just said, "Oh well, it just does. That's how it's built." I just wanted to throw that out as a question, how does the heart know? Does the mind tell the heart? Does the heart just somehow have this intrinsically? That seems like since it's electricity, it's in your field.

S CARROLL: Clearly, you want a cosmologist to answer this question. I can see why that would make sense.

I think, for one thing, we shouldn't use the vocabulary "How does the heart know." That's just like saying "If I dropped a glass, how does it know to go down?" Attributing knowledge to the glass is probably not the right way of getting the best explanation for that phenomenon. As far as I know, it is the intrinsic state of the musculature of the heart that says "All right, it's been about a second. I should beat again." I don't think it's a signal from the brain, but if there's anyone who actually knows, they should answer the question, because I'm a physicist.

Also, by the way, you can Google that.

QUESTION TWO: I'm a behavioral neurologist, and so I see a lot of patients at the interface of how problems in the actual matter of the brain affect the mind and behavior. I'm just wondering how that comes into the equation of the discussion. It seems like we are arguing that mind is something that's separate from brain matter, and yet we see these very horrible and sad stories every day of people who get a little stroke somewhere, or a little degenerative illness somewhere else, and they become a totally different person with a totally different set of behaviors and personality, etc. I'm just trying to understand how that factors into the theory of the mind and the brain?

Also, my second question is that I was thinking we heard a lot about the methods that science is using in the West, which are different from the ones in the East, which is great. But I also would like to hear what the conclusions are that came out of that research that is being done from the internal work, as opposed to observation or scientific methods?

B. ALAN WALLACE: First of all, this discussion in which I've been participating has not been a discussion about whether or not the mind exists separately from the brain. I haven't argued this point one way or the other.

I didn't come here to advocate a Buddhist worldview or affirm reincarnation. Rather, I'm suggesting that we treat the mind as being as natural as anything else in the universe. Therefore, if you want to understand it, observe it, and don't just glance at it. Find the appropriate technology for experientially exploring the mind in all its multiple dimensions, and let your assumptions fall away as you make one discovery after another.

Humanity has known for millennia that if someone gets kicked in the head by a horse, this may impair his mental abilities, something that doesn't happen if he gets kicked in the butt. Clearly, the mind and brain correlated. There are strong correlations between the mind and brain, but the fact that there is a correlation does not mean equivalence. There is no logic or empirical evidence to prove that the mind and brain are the same thing, viewed from different perspectives. It's just a reductionist belief, which is a poor substitute for scientific evidence.

The point here is not to argue for mind-body dualism, but rather to acknowledge how ignorant we are. We need to face the fact that there has been actually no progress at all in solving the hard problem of how the mind and brain are interrelated.

Marcelo commented that we don't understand the nature of consciousness. This would imply a stance of agnosticism, right? That's an expression of humility, recognizing what we don't know. But then he went a step further, declaring that no one understands consciousness. As a person who's immersed myself in Tibetan culture and wisdom for many years, this looks to me like a move from agnosticism to omniscience: since we don't understand consciousness, no one knows, and if they tell you they do, they don't know what they're talking about.

How do we know the limits of what everybody else knows? We've ignored Asia for the whole history of science. Did anybody in Asia come up with discoveries in the last five thousand years about the nature of mind and consciousness? They've been using different methods that we in the West have not used, just as scientists have been using methods they have not.

I'm not here to argue one is better than the other, but I am suggesting that it's mere hubris to say, "Nobody knows about consciousness, because we don't," we white people, we Eurocentric people. The age of colonialism should be over by now, but the remnants of that linger on in the belief that if we don't understand something, nobody does. I'm not persuaded that's true. In fact, having lived and studied with Tibetans for many years, I am quite confident there are people outside of our own culture who do have very deep insight into the nature of consciousness. They didn't get there by just believing this or that, or having great faith in the Buddha, but by doing what scientists do: rigorously investigating the phenomena themselves and seeing what comes up. Buddhists, like scientists, need to avoid ideological blockages, concluding that certain evidence can't be true, because it doesn't conform to our core beliefs. That's not scientific, it's dogmatic.

To be truly open in the twenty-first century, I think that we have to be humble with regard to the limits of our own understanding, and also humble about what other people do and do not know.

QUESTION THREE: You have two pretty different views about things, and you had a civil and respectful, curious conversation with one another. I'm curious whether either of the two of you have ever been, or are, deeply in love with someone who has the complete opposite world view of you? How do you find affection, when you know that person thinks there's life after death? You know what I mean? I swear I wanted to ask like a really intelligent question, but this is the one that kept coming to me.

S CARROLL: All right, I can talk about this. Let me talk about this. Yeah, you know, loving people is important. I think the fact that I love people is not evidence that we need to change the fundamental laws of physics, as we currently know them. But I also don't think that our knowledge of the fundamental laws of physics invalidates the way that we have of talking about human beings as conscious agents, who make choices, who have emotions, who have feelings, who have desires, who fall in love, any more that it invalidates the idea that I can talk about this chair as made of fabric and wood, even though I know it's made of atoms.

If you've ever watched *The Big Bang Theory*, you know that knowing the laws of physics doesn't make you any better at social engagements, or love, or anything like that. The reasons why you love someone, I think Alan and I will both agree, these are not things that are understood by modern science. Does Asian wisdom tell us how we fall in love with different people? I think that this is something we know a little bit about, and none of us knows completely well.

B. ALAN WALLACE: I don't know whether nobody knows it well, for that brings us again to the problem of knowing what other people do or do not know. But I think that's another topic.

QUESTION FOUR: I have this question, and this is a little more serious. I'm curious to know what happens to consciousness when the brain dies, or the body dies. Does consciousness just cease to exist, or is it conscious that something just happened, and then it just disappears. This is from your experience traveling the world and interacting with people. I'm curious to know.

M GLEISER: You'll probably get very different answers.

B. ALAN WALLACE: OK. The human mind—human states of consciousness, emotions, desires, the whole array of mental process that we experience—evidently arise in dependence upon the brain functioning. As the brain is impaired—through Alzheimer's, through brain injury, disease, and so forth—various faculties of sight, of memory, and so forth, are diminished or eliminated. Clearly, there are strong correlations between specific brain function and specific types of subjective experience.

The human mind arises in dependence upon, and is correlated to, brain function, and when the brain ceases to function, the mind that arises in dependence upon the brain ceases to exist.

If our awareness of our own minds is superficial, entailing a kind of folk psychology, or casual, unrefined awareness of what's going on in our minds, it is natural to assume that our minds consist only of the mental processes that arise in dependence upon the brain. In his book *The Big Picture*, Sean writes, "we are reflective and self-aware, with the ability to shape what it is we care about. We can, if we choose, focus our caring on making the world a better place." I share with him the importance of self-awareness, and I would add that this ability can be refined and used in ways that can lead to profound insights that rock the very foundations of our modern, physicalist view of reality. Buddhists have about a twenty-six hundred-year head start on us in this regard.

I will report on something Buddhists have discovered and rediscovered many times over hundreds of years. I offer this only as an empirical hypothesis that can be tested. I've been studying Western materialistic theories of mind for about 30 years now, and I've found that all these theories have one common quality in common: not a single one is empirically testable. They're just beliefs.

What happens if you spend 8eight, ten, twelve, hours a day probing into the nature of your mind from a first-person perspective, and as months pass, your mind—with all of its human thoughts, perceptions, memories and so forth—is calmed to the point of silence. Even though your heart keeps beating, you withdraw your awareness from your body. You withdraw it entirely into the realm of purely mental experience, which happens when you're dreaming, and you fall into deep, dreamless sleep. You let your mind shut down, in terms of all of the explicit cognitive activity, emotional and so forth. What has been discovered countless times when this occurs is that you move beyond the dimension of the psyche—the conscious and unconscious human mind, the only dimension that modern science knows about.

You move beyond the domain of the human mind, which arises in dependence upon the brain, to an underlying continuum of consciousness that's purely mental, but is not human. It has no gender, no personal history. It's a simple continuum of cognizance, a flow of consciousness. And advanced contemplatives have discovered that this dimension, and not the brain, is the repository of memories and habits.

I'm not asking anybody to believe this simply because I'm saying that it's been discovered by Buddhist and other contemplatives. Just as Sean said, you may not believe what I'm saying, so feel free to disagree. But I am saying something that is not simply conjecture or belief, but is something that's been discovered and rediscovered hundreds and hundreds of times over the last twenty-five hundred years and longer. Penetrate through your psyche, to a deeper level of consciousness, an underlying level of consciousness, that is trans-human. That's the actual repository of memories.

If we use the analogy of a computer, which comes up so often regarding the mind, the brain may be likened to the keyboard, while this subtle, underlying continuum of mental consciousness is like the hard drive. This is a hypothesis that can be put to the test of experience, but don't expect it to be easy. It generally takes thousands of hours of rigorous training to achieve such a level of meditative concentration, or samadhi. But once this dimension of consciousness has been accessed, one can try to retrieve memories from previous lives and critically check to see whether they are true and, if so, whether you could possibly have known that information from some other source.

Of course, when you try to target memories from the time before you were conceived, you may come up with nothing or with shear fantasy. But what if you come up [with] memories, and then a researcher who is open-minded, skeptical, and critical investigates to see whether those memories are veridical, or whether you have any way of knowing that information about that earlier person's life who seems to share the same continuum of consciousness as your own.

This is a theory that can be tested, and so my aspiration is to create a contemplative-scientific research facility, where people are rigorously trained for years on end in developing their faculties of attention, mindfulness, and introspection, without being brainwashed into any dogma, including physicalism. Let them explore the nature and origins of the human mind in a spirit of radical empiricism, to probe through, penetrate through this dimension of mind that is clearly human. Let them probe to deeper levels of consciousness and explore them, and see whether these alleged discoveries from the past can be replicated now. And I would like to invite open-minded scientists to participate in such cross-cultural, cross-disciplinary research, integrating the third-person methods of modern science with the first-person methods of contemplative inquiry. I'm looking to Tuscany as a place to do this. What better region than the birthplace of the Renaissance and the Scientific Revolution?

I would like to create a contemplative research facility for people who are not simply open-minded, but willing to put in the immensely hard work of ten, twelve, fourteen hours of rigorous training per day, for five, ten years, in full collaboration with psychologists, neuroscientists, physicists, and philosophers.

We need physicists, because we need to solve the measurement problem—addressing the role of the observer and quantum mechanics. Let's find out together what's true. Because we know we don't now have a solution to this problem. What are the origins of consciousness? How does it interface with the brain? What happens at death? We may have beliefs that we're strongly convinced are true, but that's no substitute for knowing reality as it is.

This is a theory that can be tested, *should* be tested. So, let's just find out what's true.

S CARROLL: Very briefly, I want to end by saying a couple of points that Alan has made more than once that I completely agree with. I think that they are important. One is this idea that there is wisdom to be found in traditions that predate the scientific tradition. I say this very directly in my book, Western and Eastern. I'm not an expert on Eastern religions or traditions, or secular traditions, but I do try to sneak them in there, into the book, whenever I can. The way that I put it was for thousands of years, the way that human beings thought about the human condition, what it was to be human, in the most careful, rigorous way they possibly could, was in the context of religious and spiritual traditions.

The idea that we learn nothing from all of that thinking seems absurd, and we should be open to that. We shouldn't accept it blindly because it's in a book, but we should certainly read some of the books and see if maybe there's something in there.

The other point of course is the importance of introspection, and seeing, learning about the mind by being the mind and thinking. I think this is also something that is absolutely worth doing. In some sense that's trivial. Of course it's worth doing, because everything is worth doing. But I would actually accept the claim that scientists can focus in too narrowly on research methodologies. Not because they're necessarily the best or the only ones, but because they're the easiest ones.

I always tell my students, "Go into cosmology if you have a short attention span, because the universe is much simpler than a frog, or the United States government." Even if you want to study the brain, it's very easy to study neuroscience by looking at what an individual neuron does.

There's a paper that literally came out, I think it was in *Cell*, that argued that neuroscientists should actually care, not just about introspection, but about behavior. To a non-neuroscientist, you go "Wait, they don't care about behavior? This is not something they care about?"

Yeah, if you can study neuron by neuron what happens, then you can convince yourself that's the only thing to do. I think we should be very, very pluralistic about where we get our information from, and no matter where it comes from, and no matter what that information says, we should be very, very careful that we check it, that we're skeptical, because that's how we build up the knowledge, piece by piece.

NOTES

1. William James, *The Principles of Psychology*, 2 vols. (New York: Dover, 1950), 2:290–91.
2. James, *The Principles of Psychology*, 2:322n.
3. Sean Carroll, *The Big Picture: On the Origins of Life, Meaning, and the Universe Itself* (New York: Dutton, 2016) Kindle, 215–219.
4. Werner Heisenberg, *Physics and Philosophy* (London: Penguin Books, 1989), 59; See also Erwin Schrödinger, *The Interpretation of Quantum Mechanics* (Woodbridge, CT: Ox Bow, 1995); Heinz-Dieter Zeh, "There Are No Quantum Jumps, Nor Are There Particles," *Physics Letters A* 172, no. 4 (1993): 189–192; Paul C.W. Davies, "Particles do not exist," in *Quantum Theory of Gravity*, ed. S. M. Christensen (New York: Adam Hilger, 1984); Michael Bitbol, *Schrödinger's Philosophy of Quantum Mechanics* (New York: Kluwer, 1995).
5. Heisenberg, *Physics and Philosophy*, 21–22.
6. John A, Wheeler, foreword to *The Anthropic Cosmological Principle* by John D Barrow and Frank J. Tipler (Oxford, UK: Oxford University Press,1996), vii-viii.
7. Max Planck, "Das Wesen der Materie" (The Nature of Matter), speech at Florence, Italy (1944), Abteilung V, Rep. 11 Planck, Nr. 1797,Archiv zur Geschichte der Max-Planck -Gesellschaft (Archives of the Max Planck Society)
8. Albert Einstein, *Ideas and Opinions* (New York: Crown, 1954), 262.
9. James Jeans, *The Mysterious Universe* (Cambridge, UK: Cambridge University Press, 1930), 137
10. Anton Zeilinger, "Why the Quantum? 'It' from 'bit'? A Participatory Universe? Three Far-reaching Challenges from John Archibald Wheeler and Their Relation to Experiment," in *Science and Ultimate Reality: Quantum Theory, Cosmology and Complexity* eds. John D. Barrow, Paul C. W. Davies, and Charles L. Harper, Jr., (Cambridge, UK: Cambridge University Press, 2004), 201–220, 218–219.
11. John A. Wheeler, "Information, Physics, Quantum: The Search for Links," in *Complexity, Entropy, and the Physics of Information* by Wojciech Hubert Zurek (Redwood City, CA: Addison-Wesley, 1990).
12. Andrei Linde, "Choose Your Own Universe," in ed. Charles L. Harper, Jr. *Spiritual Information:100 Perspectives on Science and Religion* (West Conshohocken, PA: Templeton Foundation, 2005), 139.

13. Andrei Linde, "Inflation, Quantum Cosmology and the Anthropic Principle," in *Science and Ultimate Reality*, 426–458, 451.

14. Freeman J. Dyson, "Thought-experiments in honor of John Archibald Wheeler," in *Science and Ultimate Reality*, 72–89, 88.

15. Marcelo Gleiser, "Searching for The Essence of Physical Reality," *Cosmos and Culture* (blog), *NPR*, January 19, 2011. http://www.npr.org/blogs/13.7/2011/01/19/133037010/searching -for-the-essence-of-physical-reality

16. *Udāna* 68–69: https://www.cs.princeton.edu/~rywang/berkeley/258/parable.html. For an alternate translation, see *Udāna: Exalted Utterances*, trans. Ānandajoti Bhikkhu, Revised Version 2.2, Feb. 2008, pp. 212–217. https://www.ancient-buddhist-texts.net /Texts-and-Translations/Udana/Exalted-Utterances.pdf

17. *Kālāma Sutta, Aṅguttara Nikāya* III Sutta no. 65

18. Thomas Henry Huxley and William Jay Youmans, *The Elements of Physiology and Hygiene: A Text-book for Educational Institutions*. (New York: D. Appleton, 1868), 178.

3

FUTURE OF INTELLIGENCE

Human, Machine & Extraterrestrial: A Dialogue Between
an Astronomer and a Philosopher

MGLEISER: Patricia Churchland and Jill Tarter are two wonderful pioneers, in their fields, about the future of intelligence. I should start by explaining what I mean by "future of intelligence."

Unless you've been asleep for the past five years, you know that everybody is talking about machine intelligence, machines that have, to a degree that is hard to define precisely, some level of intelligence. Machines can beat the best chess player. They can beat the best Jeopardy player. They can beat the best Go player. Are these machines intelligent? If they are not, will we one day be able to create a machine that is not just going to blindly obey commands in a program? As it is now, machines follow the lines of code we write. But can we think of a day when machines will actually start writing code on their own like a creative autonomous entity, in ways that we cannot anticipate? Is that going to be possible? And, if it will be, what will happen to us as a species?

You may have heard of Oxford University philosopher Nick Bostrom. In his book *Superintelligence*, he says something like this: "If we could develop a super intelligent machine, would we become obsolete?"

He makes an interesting parallel between us and gorillas. The future of the gorillas is completely in our hands. We could go out to Africa with machine guns and kill them all. But we haven't done that because we have some moral standard that wants to save that species. But we could. We have the power to do that. (Of course, gorillas are extremely endangered due to rampant illegal poaching. But that's another story.)

The question is, what if those intelligent machines don't have moral standards similar to ours? Are they just going to get rid of us because we are inconvenient? In other words, as we enthusiastically pursue artificial intelligence, are we also engineering our own destruction? This is an ethical question and a scientific question.

You can't—or shouldn't—just think, "Oh, let's just go on cranking and hack the hell out of the codes and do this thing," without considering the social, cultural, and moral implications of what is going on. This question, whether we should or shouldn't pursue artificial intelligence and what we can do to safeguard our future, is a perfect example of how the sciences and the humanities intersect.

And then, of course, there is alien intelligence. Can life elsewhere in the universe be smart? We don't know either way, but it is certainly a possibility since it happened here and, as you are going to find out from Jill, we now live in this amazing time where we know that pretty much every star that we see out there has planets orbiting around it.

Just pause for a second to consider that in the Milky Way alone, our galaxy, there are about 200 billion stars. Let's say that all of them, or most of them, have five planets or so. We are talking trillions of worlds just in our galaxy! And that's not including the moons, because there are planets with lots of them, like Jupiter, which has more than sixty moons. Imagine the night there. It must [be] pretty awesome. The point is that there is a lot of space. As Jill likes to say, there is a lot of real estate for life to develop, so could there be life that is more advanced then? And if so, how could we find out? And could they be listening to us as we listen to them?

A few years ago, the late Stephen Hawking scared a lot of people with his concerns about AI and aliens. First was machine intelligence. He and Elon Musk talked about machine intelligence as a possible threat to our species. And then came the aliens. We shouldn't make smoke signs because if they see them, they will come and kill us because, of course, they will be the bad guys.

And now, the latest catastrophism comes from climate change and social unrest. We may only have a hundred years to survive as a species because we are just going to kill each other as we fight for dwindling resources. This is certainly not the world I want my children and their children to live in.

There are other ways of thinking about this [that] are more optimistic and, I hope, will come up tonight during our conversation and in some of the questions, as well.

This conversation is our mirror. As we think about machine intelligence and alien intelligence, we are really thinking about our own intelligence, about who we are and want to become as a species.

What are we? At the very least, we are weird molecular machines that evolved on this particular planet and are capable of asking questions about meaning and our future and about the why and the how of things. Why we are this way remains a big mystery. As we ask questions about machines and aliens, we are actually asking about ourselves, trying to figure out how we are going to move forward from here to a future where we actually improve as a species and go beyond some of the dark parts of our history to create a society that is just and equitable.

Patricia Churchland is Professor Emerita of Philosophy at the University of California at San Diego, where she is an adjunct professor at the Salk Institute. Her research focuses on the interface between neuroscience and philosophy. Her most recent book is *Conscience: The Origins of Moral Intuition*. She is the author of the groundbreaking books *Touching A Nerve* and *Neurophilosophy*, and she co-authored, with Terry Sejnowski, *The Computational Brain*. Her book *Brain Trust: What Neuroscience Tells Us About Morality* received the PROSE prize for science. She has been president of the American Philosophical Association and the Society for Philosophy and Psychology. She won the MacArthur [Fellowship] in 1991 and the Rossi Prize for neuroscience in 2008. She was the chair of the USD Philosophy Department from 2000 to 2007.

P CHURCHLAND: Yes, there are deep and puzzling issues in front of us. I want to start talking about intelligence where I feel I am somewhat grounded and that is with respect to the biological creatures that we are familiar with—wolves, raccoons, ravens, and us. I prefer to think about intelligence, to the degree that we understand anything about it, in terms of the physical brain. For as anyone knows now, the idea that existing machines are intelligent in the way that raccoons are intelligent may be like saying that my computer mouse is like a biological mouse. It isn't.

Now, I should just say [from] the outset that we don't have anything like a definition of intelligence in the way that we have a definition of a protein or an electron. Nevertheless, there is a general sense that intelligence involves analytical, analogical, and linguistic capacities. That is basically the IQ test approach. As soon as you think about it, however, you realize that intelligence involves so much more than analytical, analogical, and linguistic capacities. It also requires what we might call practical intelligence and common sense, which is knowing what to do in varied conditions, for example in spatial navigation, social navigation, foraging, hunting, avoiding predators, having a good shelter and a safe place to hide. In other words, all those things that are necessary to surviving and thriving. We should also include as intelligent creativity in problem solving, such as how to make a boat suitable for local waters or a suitable nest if you are a bird or a chimpanzee. It concerns how to use the equipment that is your body to do what you need to do to get on in life. Because termites are very good at surviving, we distinguish the capacities of a raccoon from those of termites by referring to the flexibility to conditions of raccoons and their capacity to alter their behavior as a function of environmental contingencies. Note, however, that the aforementioned are not as easy to test as, for example, are mathematical abilities.

Social intelligence and moral or political intelligence are obviously extremely important in social mammals, and they are even more difficult to quantify. So intelligence, as we see it in biological creatures, is not a single unified thing. We all know of brilliant mathematicians who are dumb as posts when it comes to functioning in the practical world. Are they intelligent, full stop? Well, it depends on what the task is—proving a theorem or getting a clean pair of socks.

It seems probable, at this stage, that [the] cortex—that very special neural organ—is what makes mammals smarter than reptiles. Bear in mind, however, that it is even more difficult to test intelligence across species than within a species. How do you test a raccoon against a turtle, for example? We do not even know how to accurately test the intelligence of a cat against that of a racoon. We think of mammals in general, relative to reptiles, as having a greater capacity for impulse control, for long-term planning, for accessing and assessing the value of a predicted consequence and so forth. It also seems probable that members of a given species are apt to be more intelligent the larger the cortex (on average) in that species. We tend to think of humans as being especially intelligent, and we have a very large cortex relative to other primates, so it may be assumed that we have the cortex to thank for human intelligence. That is an assumption, however. So much is unknown, and when you are excited about your ideas, that may be difficult to acknowledge.

[The c]ortex is a product of mammalian evolution. All mammals have [a] cortex, from mice to rats to us; no non-mammals have it. The cortex is a darkly stained rind around the surface of the brain, unique compared to every other structure in the brain. It [is] distinguished by having six highly organized layers, with input from other areas coming to very specific places in [the] cortex, and output pathways emerging from very specific layers and going to very specific areas elsewhere in the brain.

If you look at a slice of a rat's cortex and then look at a slice of [a] human cortex under a microscope, you will see they have essentially the same organization. We don't see anything like that organization in the more ancient structures that we share with reptiles.

Here is a caution regarding [the] cortex. We know, of course, that birds such as ravens and crows and parrots and owls can be very intelligent in the sense that they can solve difficult problems and appear to have causal models of the world. Yet they do not have [a] cortex. Turns out, however, after many years of careful anatomical work to understand how they can be so smart without having a cortex, that they do have an analog of a cortex if you look at the micro details. It just doesn't have this magnificent six-layered structure that we do. And the octopus family? They turn out to show impressive intelligence in their behavior, yet they too have nothing that looks like [a] mammalian cortex.

The other thing that you should know about [the] cortex, and this will come up again in the context of artificial intelligence, is that there are very specific cell types located on very specific layers and connected [in] highly specific ways. There is nothing haphazard about the architecture. Surprisingly perhaps, different cell types respond to exactly the same input in quite different ways. They are doing different things. Even though we don't know precisely the unique function of each distinct cell type, we know that their particular shape, structure, neurochemical profile, and connectivity portfolio are crucial to how information is handled. If certain cell types end up on the wrong layers as, for example, in mice with certain genetic variants, then the animal very quickly dies of powerful seizures. Cell types and their connectivity are very important in biological brains.

And finally, of course, there is the matter of connectivity, because brain function is all about how neurons and networks are connected to each other. This slide shows a schematic of cortical neurons with little spines, and it is on the spines where contacts are made, where synapses are located. Some anatomists study the pathways that connect one region to another, and within a region, from one network to another. This is called connectomics.

We know that one of the most astonishing things about [the] cortex is that it is a big learner. All mammalian babies are born with a very immature cortex and [the] cortex gains structure as the animal learns about its world. Infant humans grow about ten million synapses per second, and their neurons sprout dendrites and axonal structure as well, which is why their heads expand during development. Much learning can be thought of as the brain constructing abstract models to reflect the animal's world, such as its spatial world. Because all mammals are social to some degree, mammalian brains make models not only of the animal's physical world, but also its social world.

Although there is vastly more to say about [the] mammalian cortex, I want to turn now to consider machine intelligence. At the outset we can acknowledge that mathematical tasks are straightforward for programmed computers, and they can do mathematical computations faster and more accurately than I can. On the other hand, it has long been also acknowledged that there are jobs that any mammal can do almost effortlessly, but that write-a-program machines can't do. Like perceptually recognize a cat when the cat is partly in shadow, when the cat has lost its tail, when it only has one ear, when the lights are bright, when the lights are dim. Regular computers with code and programs can't handle that kind of messiness. A deep question is why brains can do this so easily, whereas trying to code an algorithm for a real-world perceptual task has been a failure. One possibility is that the architecture of brains evolved in a messy world and evolved to perceive and interact with a messy world. Coding a program is not messy—it is the very opposite of messy. It is precise. That is not how biological brains operate. What is the secret to succeeding in our real, messy world?

In the 1980s, a handful of people, including Terry Sejnowski and Geoffrey Hinton, looked at [the] cortex and how it changes its structure to embody what it learns. They thought, "Look, if a rat can learn by experience to recognize different faces and places, and if that learning consists mainly [of] changing the synaptic connections between neurons, why can't we build a machine to mimic that? Let's not buy into the coding strategy; let's try a non-coding learning strategy. Like the brain." That was really the birth of the idea of artificial neural nets. So rather simple neural nets with three or five layers were built in digital computers, and indeed they could learn very simple things by being given examples. Promising though it seemed, the idea could not really come to fruition until there was sufficient computer power and the hardware to make really huge neural nets. That is, early neural nets could not handle the perception of cats either. But recent neural nets, which are really huge, can.

The basic idea is to make layers of pretend neurons, and the neurons are connected by pretend synapses whose strength could vary, up or down. The pretend neurons are just very, very simple (unlike real neurons), but the hope was that the simplicity might not undermine the project if there were enough pretend neurons. One end is the input end, the other the output end. Signals are pushed through the network from input to output. That is the basic architecture of artificial neural net in deep learning and machine learning as we now know and love it. If that is the structure, what about function? How can a network learn without a coded-up program?

The goal is to get the artificial neural network to learn, but how? The connectivity between the pretend neurons (pretend synapses) is what can change, and you want to find a configuration of connections that comes to embody the capacity to solve some task, such as to recognize a cat as a cat. Basically, you have two options to achieve this: change the strength of the pretend synapses by hand or find a suitable algorithm that will change weights (pretend synapses) in the right direction so that the final product is a network that can perform the task, such as identifying a photo of a cat as a cat.

Modifying the weights by hand to achieve your goal is not realistic, especially if you have more than about ten neurons in your network. So that leaves figuring out an automated way to do it. Finding a simple but successful way to automatically change weights took mathematical ingenuity. What worked depends on the idea that the network will get feedback to its output answer, which at first will be utterly random, but with continued feedback to answers, can get closer to being correct.

A simplified analogy for weight modification is the way we finally succeed in the game of Find the Penny. As I am searching for the penny—behind the cushions, under the table etc.—you give me an error signal so I can get closer to the goal; where the penny is hidden. "Warmer, colder, colder, little warmer, much warmer, hot." To the network's output, there is feedback—roughly like cold, warmer, hot, etc. When the algorithm feeds that evaluation back through the network, the weights get adjusted so that on the next input, the network is a tiny little bit closer to getting the right answer on the next go through. An error signal is essential to learning in baby steps, just as it is in real brains. Minus the math, that is the nub of the story of artificial intelligence—get an error signal, feed it back through the neural network causing the pretend synapses to change their weights so that they get closer to the right answer (find the penny). The weights can become stronger or weaker or maybe they stay the same.

What makes a deep learning machine deep is that it has thousands of layers, millions of pretend neurons and even more pretend synapses. To learn a pattern recognition problem, such as recognition of handwritten numerals, requires that the network get millions of input exposures to the data set and millions of error corrections. What is rather surprising is that although dinky neural networks seem rather disappointing in their capacity, if you hugely scale them up, they can be very powerful indeed. This surprised the socks off many people who wanted to create artificial intelligence by coding with a program. Artificial neural nets have no program.

At Google, under the direction of Demis Hassabis, a network was trained by examples and feedback to play the very difficult board game Go. Remarkably, when pitted against the world champion, Lee Sedol, the network won four games out of five. The wins were a dramatic achievement, without doubt, and the world was agog. Because the wins were so dramatic, they provoked people like Nick Bostrom and others to think that if a machine can learn to play Go, it can do anything that we can do. Or even more, God forbid.

Stunning as the achievement was, a host of major cautions are in order. First bear in mind that what the neural network is doing is essentially just pattern recognition. That is not nothing, but it certainly is not the sum total of what a biological intelligence can accomplish. Second, the network had to play millions of times in order to learn the patterns it could recognize as good moves. Third, if you altered just one tiny feature of the game, the network had to start from scratch to learn the game all over again in millions of trials. In that sense, its learning is rather brittle. Finally, the only thing the network can do is play Go. It cannot add and subtract, and it cannot recognize a cat as a cat. Permit me to expand on these cautions.

Among the many things that we do—and rats and monkeys and all kinds of creatures do very well—is survive and thrive and make sensible, adaptive decisions. Some of the information supporting those decisions has an external source in the environment, but some derives internally from the brain itself. One example of an internal signal is the formation of a goal such as finding food, and the goal will not be random, but suitable to the species of animal the brain is housed in. Not everything in the world is of equal interest across species. Dung beetles are highly motivated to seek dung; squirrels are not. On the other hand, squirrels are keen to find nuts and to distinguish between fresh and stale nuts, but dung beetles care not. Dogs are typically motivated to sniff the behinds of other dogs, humans are not. And so forth. Goals and plans to achieve them are internal to the animal. Often the external stimuli are essentially neutral, save how they matter to the animal's achieving its goal. Curiosity, attention, and the desire to explore are functions internal to the brain. They are not pattern recognition functions in the way that perceiving a cat is a pattern recognition function. All brains generate internal signals—desires, motives, goals, fear, need to sleep, etc.—and these internal signals play a role in what happens to such pattern recognition functions as these animals deploy.

Additionally, the biological brain constructs abstract representations from perceptual input. This is known to be true of the hippocampal structures, which construct an allocentric representation of the animal's spatial environment.

In this diagram [figure 3.1], I want to just remind you that [the] cortex, unless it is suitably connected to the rest of the subcortical structures, is sort of adrift. Unless your cortex has the appropriate sorts of connections to the deep underlying brain structures, such as the basal ganglia and midbrain, structures that regulate goals, emotions, motivations, values, and so forth, it cannot play a role in survival.

All vertebrate species can detect threats, and behave appropriately in response to motivations to survive, thrive, and reproduce. Natural selection ensures that. In this domain, as well as maintaining homeostatic functions, there are typically competing values and competing opportunities: should I mate or hide from a predator, should I eat or mate, should I fight or flee or hide, should I back down in this fight or soldier on, should I find something to drink or should I sleep, and so forth. The underlying neural circuitry for essentially all of these decisions is understood, if at all, then only in the barest outline. Moreover, these kinds of decisions do involve some sense of self, which is a brain construction, not a feat of pattern recognition. Some involve consciousness, a function which is also not itself [a] pattern recognition function. Biological evolution favors those animals whose values and decision allow them to

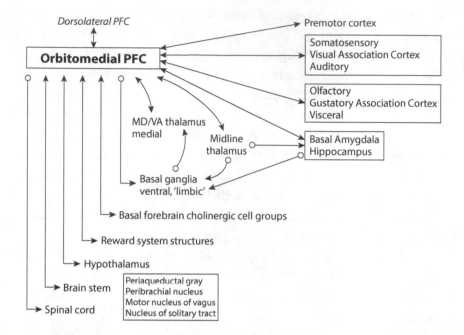

FIGURE 3.1 Schematic representation of the pathways connecting one small region of prefrontal cortex (PFC), namely orbitomedial PFC (located just above the orbits of your eyes) and a range of subcortical structures, including the hypothalamus and the reward system. Shown also are pathways connecting the orbitomedial PFC to other cortical regions. Note that most of the pathways depicted are bidirectional. For simplicity, the basal ganglia are depicted as a single area, but in reality they encompass many subregions with their own systematic connectivity. The thalamus is also a multiregional structure with its own complex pathway system.

Source: Henk J. Groenewengen and Harry B. M. Utlings, "The Prefrontal Cortex and the Integration of Sensory, Limbic, and Autonomic Information." *Progress in Brain Research* 126 (2000): 3–26.

survive long enough to pass on their genes. None of these capacities for survival and thriving do we see in artificial learning machines.

Mammals, at least, appear to build causal models of the world. Since causality is a stronger relation than correlation, the standard real-brain tactic for upgrading to causality involves intervention and manipulation. This may be easier for an animal that can move than a stationary, if deep, learning machine. The capacity for movement, especially if you have limbs, is anything but simple, that we do know.

Knowing what to test for and how to manipulate seems to be a smart business.

So far as I can tell, no one has a genuinely workable plan concerning how to capture motivation and drives and motor control into a plausible pattern recognition regime, though Yann LeCun has some ideas about happy-not-happy scales. The problem is not straightforward because motivations come in different packages—hunger is different from thirst, which is different from lust or fear or curiosity or joy. Temperament comes in different colors on a spectrum—introvert or less so, risk averse or less so, energetic or less so, and so forth. These factors change with age, with time of day, with sleep, with mood changes, and with diseases. These functions might be understood as the drivers of pattern recognition jobs in real animals, not as pattern recognition themselves.1 The hypothalamus and brainstem, which are crucial in the nervous system of real animals for the managements of these functions, are not yet well understood in neuroscience, to put it mildly. The circuitry is ancient and extremely complicated. It does not look like it is just doing pattern recognition.

What really are resilience or exuberance or patience or tenacity? Not just pattern recognition, almost certainly. How do these phenomena interact with motivation, drive, and desires? What is the role of neuromodulators in these and other phenomena such as curiosity or wanderlust or empathy or sociality or aggression? Neuroscientists are indeed exploring these phenomena, one and all, but their neurobiological mechanisms are not easy to plumb. That they are simply, at bottom, forms of pattern recognition seems unlikely at this point.

Go ahead and market an artificial neural network as intelligent, but if it is brittle, lacks flexibility and common sense and has nothing approximating motivation or drive or emotions or temperament, it may be difficult to persuade the rest of us that it is intelligent in the way that biological entities can be. Redefine intelligence, you may, but the redefinition per se will not make the machine intelligent in any generally recognizable sense.

At least some of the dystopian predictions concerning the eventual threat to humans of intelligent machines depend on the false assumption that engineers have now cracked the problem of intelligence in a machine. However dramatic such predictions may be, they are not tethered by a biological understanding of what makes for intelligence, and they certainly are not grounded in a biological understanding of the nature of motivation and goals. Waving your hands excitedly is not really a compelling research strategy.

Francis Crick used to say that you can make a realistic prediction in science only about five to ten years out. That is the prediction horizon. After that you are just indulging in wanton speculation. While such speculating may be oodles of fun, it provides no reason for the rest of us to reorder our lives accordingly. I tend to feel that much of the commotion about the threat of artificial intelligence is way beyond the prediction horizon. While it has produced some insights, it has also produced a lot of attention-getting nonsense. I am not indifferent to the human plight, and AI as it stands will certainly have enormous economic impact. But as for fearing that the machines will do away with humans as they take control of us, I prefer not to waste my neural activity on a problem that is unrealistic, especially when realistic problems are staring us right in the face. Climate changes, species extinction, and a pandemic, for example.

M GLEISER: Jill Tarter holds the Bernard M. Oliver Chair for SETI (Search for Extraterrestrial Intelligence) Research at the SETI Institute in Mountain View, California, and serves as a member of the board there. Tarter received her Bachelor of Engineering Physics at Cornell University and a master's and a PhD in Astronomy at the University of California. She has spent the majority of her professional career attempting to answer the age-old and essential question, "Are we alone?"—the question we all want to know the answer to—by searching for evidence of technological civilizations beyond Earth.

Since the termination of funding for NASA's SETI program in 1993, she has served in leadership roles to design the Allen Telescope Array and to secure private funding to continue the exploratory science of SETI. If you want to be participatory on this, fund SETI. Many people are now familiar with her work, as portrayed by Jodie Foster in the movie *Contact*, but more to the point, there is a recently published authorized biography of Jill Tarter by science writer Sarah Scoles called *Making Contact*.

J TARTER: Patricia gave us a wonderful exposition about human and machine intelligence. What about the extraterrestrial? I am going to try to change your perspective and I want to make you think about this fact, that you are made of stardust. Quite literally. It takes a cosmos to make a human. The iron in the hemoglobin in your blood, the calcium in your bones and in your teeth, they were all cooked up in the interior of massive stars that blew up billions of years ago. In fact, our story, the story of humans, began billions of years ago.

Our universe was born in an immensely energetic dense hot explosion something like 13.8 billion years ago. Our galaxy was born 10 billion years ago, and we humans actually trace our lineage, not just back through the centuries of our families, not just back through the millennia of our civilization, its buildings, its arts and its many experiments with governance. Not just back the millions of years since we branched off from the great apes, not just back but 2.4 billion years during which our atmosphere has been perfused with oxygen thanks to the photosynthetic labors of cyanobacteria.

Not just back to the 4.6 billion years ago when our sun and planets were born, but back another four or five billion years when giant molecular clouds were contaminated with the winds of Wolf Rayet stars and supernova explosion detritus. The ends of life of massive stars, supernova explosions, leaving remnants such as this modern remnant. The stuff that got incorporated into the stars, the planets, to life, to us, and which may happen again in the future to this modern remnant. We are made of stardust.

It has taken millennia to piece together this scientific story of where we came from and we continue down this road of inquiry because we are curious. We want to know more, we want to know why we are here and how we fit in and what else is out there and, of course, is anybody else there? As we journey down this road, people, my colleagues, astronomers and astrobiologists, have a different perspective on what it really means when we say here and now.

This is a much broader cosmic perspective and I think it is something that we should spend more time sharing with the public, with people who are our leaders, for our near and long-term future. We need to share this perspective because it is important, it matters to our long-term future. Here, it is simple right? We are here in the Feinberg Theater, within the Venue SIX10, in Chicago, Illinois, and we are all now pretty much around the world, comfortable with this kind of Google Earth perspective.

But from the altitude of low-Earth orbit satellites, we see ourselves as here. Since 1968. when Bill Anders took this Earthrise picture as he was coming around the limb of the moon, we have been able to see ourselves from this perspective. This is our first truly cosmic perspective. Seeing ourselves as a single entity floating in space. OK, actually, in 2013, the *Cassini* spacecraft, which has been orbiting Saturn, turned around and looked at all of us. They sent out a memo—I hope you went out on the lawn and waved—when Cassini took this selfie.

There is a white dot there and that is us, all of us. It was done earlier in 1990 when the *Voyager I* spacecraft, going past Neptune, turned around, looked homeward, and saw this pale blue dot. That is all of us. And our sun is only one of 400 billion stars in a beautiful spiral galaxy like this one. Now, nobody got a spacecraft out there to turn back and look at this; this is another galaxy, which is analogous to our own. We are not even anywhere near the center, we are out in the boondocks.

And that galaxy, our Milky Way, is one of 200 billion other galaxies in the observable universe. Those white dots, they are not stars. Each one of them is a galaxy with hundreds of billions of stars. As we look at these beautiful images from the Hubble Space Telescope and notice that some of those galaxies are smaller and fainter, we are reminded that as we look farther out in space, we are looking farther back in time because of the finite speed of light.

And so, we have to encompass now as being the current epoch of evolution of a universe that started about 13.8 billion years ago and has gone from the big bang to big brains and is going beyond. We have to encompass this sense of time and our place within it.

Consider two images. One, an image of a neuron. Patricia was just talking about this. It is only a few microns across The other, a "baby" picture. Simulated, because no one was there to take it, but a baby picture of our universe showing the birth of galaxies in the knotty tendrils of the cosmic web.

Do the similarities in these structures tell us anything about what to expect beyond Earth when we go looking for intelligence? Well, here is the Milky Way from Earth and one thing that has changed profoundly over my career is that we can now tell you, with very great statistical certainty, that there are more planets than stars out there. There is potentially a lot of habitable real estate and now is the time to try and find out if any of it is actually inhabited.

My astrobiology colleagues are very eager to study the atmospheres, if any, of the nearby exoplanets. Looking for biosignatures, looking for something that might affect the chemistry of the atmosphere in ways that only biology can. They are looking for microbes. I am interested in finding mathematicians and, to do that, we have to use a search that looks for technosignatures. Although we call it the search for extraterrestrial intelligence, we are actually using technology as our proxy for intelligence.

That is all we know how to look for and whether or not this kind of search will be successful is completely determined by the average longevity of a technological civilization. Because unless technologies persist for a long time, there will never be any two close enough in space and lined up in this deep temporal history in our galaxy to interact and discover each other. Phillip Morrison has a lovely way of stating that. He says that SETI is like the archeology of our future.

If we ever detect a signal, and there is information, it will be telling us about their past, because this tyranny of light speed means the signal has been traveling a long time to get here. If we detect a signal, we know that it is possible to survive our technological adolescence. We know that it is possible to become an old technology. Someone else made it through and we can as well.

We have been doing this SETI for about fifty years and, let me tell you, we've worked really hard. But if you take the nine-dimensional volume that we need to search for signals and you set that volume equal to the volume of the Earth's oceans, to date we have sampled one glass, one twelve ounce glass, of the earth's oceans. Maybe it is not surprising that we haven't found anything yet, but what is very exciting is that we are, right now, at a point where we can grab the exponential expansion of our computing capacity and our ability to build bigger glasses, to put them in the ocean faster and to search much more rapidly.

In Northern California, thanks to the generosity of Paul Allen, we built the Allen Telescope Array up near Mount Lassen. It currently has 42 dishes, and they are all hooked together by computing. Silicon is as precious to us as aluminum and steel. The wonderful potential for this array is not only to build more dishes, maybe as many as 350, for more sensitivity, but to grab onto Moore's law of exponential growth to be able to detect much more complex and many more signals at one time.

In the optical part of the spectrum, we began doing SETI [at] about the beginning of the twenty-first century, and here is a dedicated telescope built at Harvard University, which searches the sky looking for very bright, very short optical pulses. The kind of thing that a laser can do, but mother nature can't. Over the years, they have scanned the sky that can be seen from Massachusetts a number of times.

More recently, Yuri Milner has generously promised a pledge of 100 million dollars over the next ten years for additional SETI searches and a project called Breakthrough Listen, where they will rent time on large astronomical telescopes and build backend signal processors. Right now, they are at Green Bank, West Virginia; Parkes, Australia; the automated planet finder telescope at Lick Observatory; and, soon, China's FAST telescope and the MeerKAT telescope in South Africa.

Now, we are looking for artifacts, we are looking for things that look engineered. In 1973, Arthur Clarke told us about the third law, Clarke's third law. He said, "Any sufficiently advanced technology will be indistinguishable from magic." We have been imagining what that magic might be. We have been projecting forward our technology and looking for these kinds of artifacts. More recently, philosopher Karl Schroeder has said a really old technological civilization might, in fact, be indistinguishable from nature because that is the only way that you can get to be old.

As we look at exoplanets for biosignatures, we might look for strangely engineered, over-natural planets, and that is really hard to figure out what I mean by that, or what anybody means by that. Maybe they artificially change their climate so there is uniformity in latitude. You don't have cold poles and warm equators, they might get rid of extreme weather over time, and, ultimately, they might modify their star in ways that give them the energy they need to do these things, to keep their world in homeostasis and natural.

Of course, then what would an engineered, natural world look like if the world were run by machines? This is one of Nick Bostrom's nightmares, that we take insufficient care in building a super intelligent machine. We don't constrain and box it with enough security. We don't take enough time to think about a good goal, a moral goal, the right goal to give a machine. We might tell it to build a million paperclips, in which case the machine turns everything available into paperclips and paperclip manufacturing facilities.

We don't know what such a world would look like as we go looking and we wonder, is this our future as well? Right now, when you don't know how to define what you are looking for, machine learning and artificial intelligence are beginning to give us, with these neural nets, some new tools. We are going to try and do this in SETI, use machine learning to find patterns in noise that we currently are not searching for [and] aren't aware are there. We have been working with IBM for the past eighteen months or so and we are going to be holding a hackathon in San Francisco on June tenth and eleventh, and follow that with a summer-long code challenge. So if any of you are interested in this kind of thing, please think about contacting us, going to that website and registering.

I am just going to end with one last sentence from Caleb Scharf, the chairman of the astrobiology department at Columbia University. Caleb reminds us that, "On a finite world, a cosmic perspective is not a luxury. It is a necessity." That is the point of view that we all need to adopt in order to be able to manage and find solutions to those challenges that look so difficult now, particularly the challenges that don't respect national borders.

M GLEISER: Now we have time for conversation. I will just ask a couple of questions and then feel free to ask questions to each other as well. Let's start with Patricia. The other day, I was landing in San Francisco and was taking a Lyft from the airport to town, and there were these billboards about artificial intelligence and machine intelligence every-where, as they are already selling it as a product. I wonder if that bothers you? I think calling our current technology AI is a misnomer.

P CHURCHLAND: Well, I don't think it really matters if we call it artificial intelligence. Doing really complex pattern recognition is a kind of intelligence, it is a tremendously important kind of intelligence, and I think the machine learning, deep learning, is going to have tremendous numbers of uses, without a shadow of a doubt. It is already being used in your smartphone, of course, and that is what Alexa uses to do voice rec-ognition. It is being used in medicine to read mammograms and other forms of radiographs.

It may be useful to remember that it is a very, very restricted kind of intelligence and it is not the kind of intelligence that is going to lead us to the Bostrom catastrophe because to get from here to there, you need a huge amount of other stuff—motivation, goals, capacity for causal inferences, desire for control, and so on. Additionally, you need to have devices that can learn [from] one or two trials, not many millions.

As noted earlier, if you change one tiny rule in a very minor way in the game of Go, poor old Alpha Go has to start from scratch. I can teach a dog to make one tiny change in how it herds sheep; it will learn in two trials. It is also the case that each one of these deep learning machines can do one pattern recognition problem. It can recognize faces or it can recognize numerals, but it can't do the many things that a sheepdog can do. It can't herd sheep, [and also] bring in the paper, recognize faces, and learn not to dig holes. The AI machine can do one thing. I don't want to diminish that achievement, it is a very, very important technological development, but people kind of talk themselves into attributing much more to the neural network than it genuinely has.

M GLEISER: I want to go back to that, but before we do, Jill, could you tell people a little bit about the Fermi paradox, because it is something that is all over the place and people hear about it and there are a lot of expectations and potential explanations. I wonder if you can say some-thing about it and what your favorite resolution is.

J TARTER: Well, paradoxes in astronomy have a long history, they can be extremely powerful. Observing Olbers' paradox, observing that the sky is dark at night, finally leads us to understand that the universe is expanding. Paradoxes can be very powerful structures, so people take the Fermi paradox seriously. It has its origins in Enrico Fermi coming to lunch one day at Los Alamos National Lab and saying, "So where is everybody?" His colleagues understood that what he was doing was constructing a story, which says that it takes, cosmically speaking, such a short time to go from the origin of technology to the origin of space flight and the ability to move among the stars, that if there had ever been anywhere, any when, another technological civilization, then in times that are short compared to cosmic times, this previous civilization would have occupied the entire galaxy. They would be here. You can make up all kinds of different models, it doesn't matter, to an astronomical accuracy. They would be here in a very short time. "But they are not here. That must mean that there could never have been another civilization with technology. That we are the first."

You have got to take Fermi seriously, right? This was an interesting proposition. But I say that it breaks down to a logical, philosophical paradoxical structure, because we cannot say that they are not here. I don't mean that they are abducting Aunt Alice off the streets of Chicago for salacious medical kinds of experiments. I don't mean Bigfoot, I don't mean flying saucers, but go back to the how big here is and how deep now is and we have explored, even our local solar system, so poorly that we would have missed any number of Coke cans filled with artificial intelligence that was, in fact, them. We can't even guarantee that we have found all of the big rocks with our name on [them] and we have got a real vested interest in that one. I think that we overestimate when we say that we have looked and they are not here.

P CHURCHLAND: Well, I am sort of curious. When you think about what the intelligences might be, they might be just really, radically different from us. They might be kind of big, smart gasbags or something.

J TARTER: A really good science fiction novel called *The Black Cloud* imagined just that.

P CHURCHLAND: What do you think they might be like, assuming that you are right, that it's very probable that somewhere out there are intelligences of some kind?

J TARTER: I don't . . . I can't answer that question. I look at where we came from, the building blocks that made us, that made your rats and your sheepdogs and the microbes and might that process be repeated? Is intelligence actually something that will be selected for evolution-arily? It is pretty expansive. On the other hand, if there is any predator/prey relationship on some other planet, evolution may ratchet that up pretty quickly and maybe you get intelligence. And I think that it is not unreasonable to think that there is something out there that can build and operate a transmitter. Which, of course, is my cop-out on trying to define intelligence. It can be a big gasbag if, somehow, it can manipulate the physical world in order to produce evidence of its existence over interstellar distances.

P CHURCHLAND: It might mean a bunch of dolphins who just don't have these electronic devices. They have got fins and they can blow these wonderful bubbles and they can cooperate with each other, but they don't have the technology because they do not have the physical equip-ment to produce it.

J TARTER: That is right, and they may be able to write poetry and sing songs, but we wouldn't know it from here. The other thing they don't have is the ability to make fire and that may indeed be necessary for technology. It may have to be land based, not water based.

M GLEISER: Right, because you need to be able to manipulate tools in order to develop technology.

P CHURCHLAND: Yes indeed. For hominins, fire was probably cru-cial to our cortical expansion, perhaps because we could cook and get more calories out of our food, as Richard Wrangham has proposed. That allowed more time for doing something other than foraging. Making better tools, for example. And getting metal out of dirt.

J TARTER: That would be a wonderful world to find, a world of dolphin philosophers. [laughter] No kidding, but I don't know how to find it. I don't know.

P CHURCHLAND: I would like that, too.

M GLEISER: If you look at the history of life on Earth, it is kind of an amazing history if you think about it. One thing that is important for people to keep in mind is that the history of life on a planet mirrors the planet's life history, in the sense that what happened here will not happen elsewhere ever. Because the history of this other planet is necessarily going to be different: even if there are other Earth-like planets out there, they are not going to go through the same events that happened here, all the impacts and global cataclysms and so forth. As a consequence, one thing that we can say for sure is that we are the only humans in the universe. Not a bad thing to consider when you are feeling a little blue.

J TARTER: I think that is absolutely correct, and the other thing to remember is that the planet has shaped the life that is on it, but the life on this planet has also profoundly shaped the planet. If there is life elsewhere, I think there will be this co-evolutionary development. We have no idea how many different pathways there might be to creating something, again, that can build and operate my telescope.

M GLEISER: Going back to the artificial intelligence issue, Patricia, can you tell us a little bit about the relationship between the nature of intelligence and free will? I know this is a hard question maybe, but if there is one person who can answer it, it is you.

P CHURCHLAND: Hard to do briefly. Well, there is one way you can think about it in a small package. That what happens as a result of the emergence of cortex in mammals is that there is a kind of slack introduced with respect to the genetic control. The cortex enables evaluation and comparison of goals and consequences, along with suppression of maladaptive behaviors. This gives the brain much more flexibility in decision and behavior than exists in, say, bees and termites. There is, if you like, a kind of buffer between the decisions that are made by [the] cortex in conjunction with the more ancient structures. The behavior of reptiles and insects is much more closely linked to genetic control.

Because [the] cortex is such a powerful learner, it means that humans and other mammals can acquire certain sets of skills and habits that are context sensitive. This is possible because the reward system, which is ancient, interacts with the cortex in such a way that the brain acquires greater and greater impulse control and the capacity to see longer and longer into the future and to evaluate consequences of an action, and hence to guide behavior accordingly. Self-control is very impressive in mammals, perhaps the more so the larger the cortex. Without self-control mechanisms, flexibility and choice are highly limited. Free will, you might say, is dependent on self-control, causal knowledge, and the ability to predict the consequences of an action.

I think that is as good as free will gets. What we have in humans is not different in kind [from] what you would have in a chimpanzee or a monkey, but different in degree. Human decision-making involves a little bit more in the way of the capacity for impulse control or to deliberate and evaluate before making a decision. We see those things, albeit [to] a lesser degree, in dogs and in rats. There are wonderful experiments showing that rats can defer gratification very nicely. At least most of them can, though there are some that are much more impulsive.

J TARTER: The same is true for humans.

P CHURCHLAND: The same is true for humans, absolutely. I think that this looseness of the connection between genetically dominated reflexes and the cortex, which can interact with those ancient structures and modify the nature of those reflexes, that is what is really kind of amazing about the development of [the] cortex. It all came about because animals became warm blooded. Once they became warm blooded, they needed to be smart because they had to eat ten times as much as their cold-blooded friends.

As soon as you have an ecological constraint like that, that you have got to eat ten times as much, then mother nature, so to speak, saw to it that there were changes made so that certain animals had this amazing structure interlaced with the ancient structures that could learn a lot.

J TARTER: Where do fire and cooking food come into that?

P CHURCHLAND: Oh, well it is very interesting about fire. As suggested earlier, fire may have been useful for cooking and extracting more calories from food, but Homo sapiens (which we are), have been on the planet for about three hundred thousand years, but there were, of course, hominins long before that. We overlapped with Neanderthal for about thirty thousand years and there was, indeed, hanky panky. I have about 4.9 percent Neanderthal genes.

J TARTER: That is higher than average.

P CHURCHLAND: Apparently so. Homo erectus appeared on the planet about 1.8 million years ago. They were small, they had a brain of about eight hundred cubic centimeters. For comparison, your brain is about fifteen hundred cubic centimeters, so it was almost half your size, but *Homo erectus* had fire and they knew how to use it. They presumably cooked their food, they had a variety of stone tools, and they appear to have made rafts by lashing logs together with vines and got themselves out of Africa into Indonesia.

Before that, there were the Australopithecines, who were rather different. They were not quite as bipedally adept and they had other physical differences from later hominins. We don't think they had fire, but every year the archeological evidence comes in and we find that further and further back is evidence of really quite sophisticated tools, not only from stone, but also antlers and bone. *Homo sapiens* are not the only ones, it now appears, that had language. Almost certainly, the Neanderthals had some language use. They probably did not have the ability to produce as many vowels as we do, but you can get along with fewer vowels than we've got. In some instances, *Homo neanderthalensis* produced cave art and buried their dead. Our last common ancestor with chimpanzees was about five to eight million years ago. Although chimpanzee brains did not greatly expand since then, in our branch, brains expanded rather rapidly.

M GLEISER: Jill, you said something before, that if there are intelligent aliens, they would have to be kind and somewhat wise. Can you revisit that argument, because I think that is a very important one for everyone to learn here?

J TARTER: That's the, "Why is a nice girl like you taking a contrary point to Stephen Hawking?" question.

M GLEISER: I am with you. I think Stephen Hawking needs Prozac or something. His views of late are a little depressing.

J TARTER: We are a young technology in a really old galaxy and we, frankly, don't know whether it is going to be possible to grow old. If you take the point that there are advanced technologies out there, advanced enough to be able to come and be a threat to us, the Hawking question, then they are going to be older. They are more advanced; it took them time to get there. I think the fact that they are old enough to have that technology, perhaps, argues for the fact that they have had to shed, get over, get rid of some of the aggressions that helped get them smart in the first place.

That cultural evolution might, in fact, lead you to become less aggressive when you figure out that you have got one planet and you have got to manage it with a certain number of individuals and manage it sustainably for a long future. I think, perhaps, there is something in that scenario that says you can't get old, you can't get that job done, unless you are kinder and gentler. It is a contrary point of view. Stephen may be right, they may all be sons of bitches and they are going to land here and we are not going to like it. But guess what? If they can land here, it doesn't matter. They are going to write the rulebook.

M GLEISER: Right. Here we are, we are the only humans and we have this very special small pale blue dot. This century, I think, is the defining century for us in many ways. All of this new science is, I think, forcing us to grow up, because that is the only hope we have. That is something that everybody should take to heart and really try to make a change. We don't have much time. You don't have to wait for the government or for corporations to make things happen. We can make things happen ourselves. Small individual changes in lifestyle can yield huge results, like a contagion for the collective good. With that, we are at minus-sixteen seconds, and so I will end this conversation and open it up to you guys in the audience to ask questions. Apparently, there are microphones that are going to circulate, and you can ask questions that way.

M GLEISER: And now we have some questions.

QUESTION ONE: I have a question about Mars and going to Mars. It seems that it may be possible within ten, fifteen years. How do you think that is going to impact the part of science that you are working on?

J TARTER: Well, let me start by answering that I think before we go there, we have to figure out if there are any Martians there, because they might not like having us show up and ruin their planet for them. This is a serious question. It sounded like a jokey answer, but it is a serious question about searching for life beyond Earth. Mars may, in fact, have evidence of fossil life or even extant life today beneath the surface.

There is water somewhere on Mars, there is liquid water, and where there is water, there may be life. We are very eager to explore Mars, to find out if there is a second genesis. Is there life on Mars and, if there is, is it related to us? Early in the history of the planetary system, Mars and Earth and Venus were all swapping rocks and a microbe could actually have hitched a ride from Mars to Earth, seeded this planet, and so it might be that we are all Martians. Could be.

Or, perhaps more interesting even, showing my bias, is that we could find life on Mars and it could be unrelated to us. All life as we know on this planet is closely related. We might find an independent second genesis of life on Mars and that would tell us that life is ubiquitous, it is going to be everywhere out there if it happened twice in this one solar system. First of all, I would like to check if there are any natives before we start thinking about exploring and, perhaps, changing the environment of the planet to be one that is more suitable to life as we know it. I am excited about the possibility for exploring and finding out what is there.

QUESTION TWO: It was mentioned that computer learning and technology only do one thing, say playing Go or playing chess or, say, a language program. Language programs are pretty good, they can even do speech now. If you look at computers and that sort of information technology, we are only maybe fifty or a hundred years into it at most. The fact that we already have language programs to me means in another hundred years or another ten years, it might get much, much better at a much accelerated rate of performing functions. Why do you think that that won't happen, or they won't be able to do more than one thing?

P CHURCHLAND: Well, I don't want to say that I think it can't happen and maybe in a hundred years something like that will happen. My point is really a more practical one, which is that certain limitations that we recognize quite well right now in learning machines we don't exactly see how to overcome. It may be that once we know more about natural, biological brains, we will see how to overcome those.

If we take Francis Crick's rather practical take on these things, he used to say things like, "You really don't want to get too excited about what might happen fifty years from now unless you are pretty sure that you can see a well-drawn line to that." We can't in the case of deep learning. We don't know how to overcome some of the important issues having to do with goals and motivation. I know it is easy to wave your hands and say, "Oh well, we will just make them smarter."

There are really significant technical limitations on that. We don't, for example, know whether you need very different cell types in order to do something like what brains do in the way that, as I showed you, [the] cortex has these very different cell types ordered in these highly canonical ways. Every unit in a machine learning device is the same as every other unit and it doesn't matter at all whether it is way at the top or in the middle or down at the beginning. Boy oh boy, in biological brains, it makes all the difference. There are many technical issues and it is, I think, just a little bit facile, if I may be rude, for Nick Bostrom to say, "Oh, there are going to be these superintelligences and we [had] better all watch out." Oh, yeah right, get on with it.

J TARTER: Actually, Patricia, one of the things that seems to me that is pretty soon going to be a stumbling block is the fact that we actually don't understand how these neural network black boxes really work and we don't really understand what biases they have built into them themselves or that we put in initially.

P CHURCHLAND: It is true that there is much about the mathematics of big networks that we don't understand. Dinky networks, if they are two or three layers, we pretty much have that nailed, but you are absolutely right. If you are looking at a learning machine that has hundreds of layers, millions of pretend neurons and billions of pretend synapses, we actually do not understand the mathematics and that is one thing that mathematicians are quite excited about because it is a job that they think they can solve. They may well, but it is true that we don't currently have a sound grip on the mathematics of all that.

In the deeper sense, with regard to biological brains, there are fundamental things that we don't understand about why brains are the way they are and how they work and how it is possible for a raven to learn, in one trial, how to solve a particular, difficult, and novel problem. No machine learning device can do what the raven can do. Why is that? Some of these limitations may be overcome once we get real neural net chips. You may scale its learning trial runs down to only hundreds of thousands instead of millions, but that is still very different from what biological organisms can do. There is something we have not yet nailed about mammalian learning.

M GLEISER: Yes, it is not just an engineering problem.

P CHURCHLAND: Well, we don't know. I don't think we know what kind of problem it is. It is partly a neurobiological problem. It is partly, I think, that we lack certain conceptual tools at the moment to know what brains really are up to.

M GLEISER: The million-dollar question, I guess, is would these machines, which could be intelligent, be conscious?

P CHURCHLAND: Oh, I have no idea, because we do not yet understand the mechanisms for consciousness in biological brains of mammals. But, as remarked earlier, consciousness is not pattern recognition.

M GLEISER: Exactly. That is a complicated question, right? They will be smart, they will do many things, they are much smarter than we are in many ways.

P CHURCHLAND: No, well they are not though.

M GLEISER: Well, they are fast at calculating and we cannot beat them, right?

P CHURCHLAND: I have no problem turning off my computer.

M GLEISER: I would like to know how many here would not have this problem.

P CHURCHLAND: Yes, I don't know. I am very interested in the problem of the nature of consciousness, and, unlike my friend David Chalmers, I do think it is a function of the biology. He thinks it is a fundamental property of the universe, along with mass and charge. I think it emerges from the biology, but there are many things about the brain that we do not understand. We were talking about this before we came on.

We understand a certain amount concerning how brains change when they learn something. If I ask you now, do you remember your first kiss? I bet most of you do. We have no idea how that memory was retrieved. Absolutely none. Now I can say, "I think recollection is probably a fundamental property of the universe along with mass and charge. It just kind of gets right in there." I am not likely to do that. Similarly with consciousness. If you can affect consciousness with LSD and propofol and a blow on the head and if it vanishes when you go to sleep and it emerges again when you wake up, it smells biological to me.

M GLEISER: It is still biological. That doesn't mean it is easy to understand.

P CHURCHLAND: No, it doesn't mean that it is easy to understand, but that does not imply at all that it is like mass and charge.

M GLEISER: Right. Maybe we need a perspective from a different kind of intelligence. That will do it because then we'll have two things to compare it to.

J TARTER: When you only have an example of one, it is just damned difficult to figure out what is necessary and what is contingent. It is only when you get multiple examples that you can figure out, "Oh, well you could have done it this way, you could have done it this way."

QUESTION THREE: My question speaks to what Patricia said about how much we don't understand about the brain already. We know that there is a very small percentage of our brains that we actually use and there's a sufficient amount of evidence, with monks changing their body temperature just by thinking and things like that. How much does all the stuff that we don't know about the brain, that huge amount of potential that we haven't tapped into, play into how you guys think about this? I hear a lot of perspectives on the way that we currently understand ourselves and our world, and what you would expect for other consciousnesses, what other beings would need to use in order to have tools or fire or whatever we think we need to have in order to survive. But how much does what we don't know impact how you think about that versus not having certain assumptions going into it?

P CHURCHLAND: It is interesting. Let me say one thing first and that is I think that, at any one time, you are not using every neuron in your brain; that is true. Probably over time, every neuron in your brain has a job. It is not like you can just get rid of some and nothing would happen. That is more true even of the subcortical structures, all of which seem to be really important. You can lose through stroke, for example, a chunk of cortex and you will still be very functional even though you may have a somewhat reduced capacity to, say, recall autobiographical events. Were you to lose even a tiny amount in your thalamus, you are hosed; if you lose a region of your brain stem, then you lose consciousness and perhaps control of heart and lungs. You are done.

Over the last three or so decades, we have learned a lot about individual neurons and how they work. They are not simple, as they are in learning machines. They are amazingly complicated. They are kind of computational devices in their own right. We have also learned quite a lot at the systems level, using things like imaging and scans and so forth, as well as various data from psychology.

The organizational structures between systems and cells are neural networks. There are macro networks and micro networks and all of that connectivity. In our current understanding in neuroscience, networks are where the big, big puzzles are. We don't know how individual neurons work with others to process information or to retrieve a memory or to collaborate on making a decision about this long-term plan rather than that long-term plan, to say this rather than that. We have essentially no idea how neural networks work. One hope, and I think this is a fruitful possibility, is that there will be a collaboration between machine learning research (the artificial neural network research), and biological neural network research so that we can have a better understanding of how brains actually do things like learn and remember, not to mention decide and generate goals.

M GLEISER: Perhaps the future of intelligence is not so much out there, but it is in here, through a connection between us and future technologies that will just be enhancers of our biological and physiological abilities. Cyborgs!

P CHURCHLAND: Machine-brain interface is definitely in the cards and I think it is really spectacular to see some of the progress that has been made in devising ways to enable paraplegics to use their brains to move a cursor, and to communicate, in some instances, for the first time in ten years. I think that is the thin edge of the wedge and I have tremendous hope for that because there are many, many people who have disabilities where this would be an absolute boon.

J TARTER: IA as opposed to AI? Intelligent Augmentation.

NOTE

1. Terry Sejnowski, Howard Poizner, Gary Lynch, Sergi Gepshtein, and Ralph Greenspan, "Prospective Optimization," *Proceedings of the IEEE*, 102, no. 2 (2014): 799–811.

4

THE NATURE OF SPIRITUALITY

A Dialogue on Science and Religion

MGLEISER: Rebecca Goldstein is a very acclaimed philosopher and novelist. She's the recipient of numerous prizes for her fiction and scholarship, including a Guggenheim Fellowship and a MacArthur [Fellowship]. In 2012, she was named a Humanist of the Year by the American Humanist Association. In 2015, she received the National Humanities medal from President Obama. She is the author of ten books, including *Plato at the Googleplex: Why Philosophy Won't Go Away* and a wonderful novel called *Thirty-Six Arguments for the Existence of God: A Work of Fiction*. She is currently a visiting professor at the New York University as well.

Alan Lightman is a physicist, a novelist and an essayist—the first at MIT to receive dual appointments in the sciences and the humanities. I'm very jealous. He's the author of five novels, two collections of essays, a book-length narrative poem, and many nonfiction books on science. His novel *Einstein's Dreams* was an international bestseller and published in thirty languages. His most recent books include *Screening Room: A Memoir of the South*, one of the *Washington Post*'s 2015 best books of the year, and *The Accidental Universe*. His latest book is *Searching for Stars on an Island in Maine*, an inspiring and personal meditation on science and spirituality. On top of all this, Alan has received the gold medal for humanitarian service to Cambodia awarded by the government of Cambodia, celebrating the work of his Harpswell Foundation, a nonprofit with a mission to equip young women in Southeast Asia with leadership and critical thinking skills.

Both Alan and Rebecca are fellows of the American Academy of Arts and Sciences. They embody in their work and in their personae the mission of bringing together the sciences and the humanities to address the big questions of our times. Their topic is the nature of faith and the relationship between science and religion. The starting point of this conversation about science and religion is that it should be a conversation and not an antagonistic kind of conflict, as so many people portray it. The objectivity of science versus the intangibility of faith. However, if you think seriously about science and religion, you're going to find out that there is an element of subjectivity in both. They both address, in their own ways, our existential urge to know the world, to understand how we fit in the big scheme of things, to understand who we are and what the purpose of our lives is.

Religions have been answering, or trying to answer, these questions for millennia and, for the past four hundred years or so, science has been trying to do the same. But it is now, in the twenty-first century, as science has advanced to address fundamental questions of origins, that we can see this confluence of ways of knowing. Questions that were once the province of religion are now also scientific questions. For example, the origin of the universe, the origin of life, the nature of consciousness, the future of humanity in considering ethical decisions about artificial intelligence, the growing automation in the workforce, these very fundamental issues that we are confronting right now force us to sit together to consider the contributions from scientists and humanists. My firm belief is that way forward requires a multidimensional approach that integrates different ways of knowing.

A LIGHTMAN: For many years, my wife and I have spent our summers on a small island in Maine. It's only about thirty acres. There are no ferries or bridges to the island, so everybody has to have their own boat. My story concerns a particular summer night, very late at night, when I was coming back to the island alone in my boat, and I just had rounded the corner of the island. It was a moonless night and very quiet, and the

stars were vibrating in the sky. I took a chance and turned off the running lights of the boat and then I turned off the engine, and I just laid
down in the boat and looked up at the sky. Looking up at the stars from
the ocean on a dark night is a mystical experience.

After a few minutes, my world had dissolved into that sky. The boat
disappeared, my body disappeared. I felt like I was falling into infinity.
I felt an overwhelming connection to the stars, and the vast expanse of
time extending from the infinite past before I was born to the infinite
future after I would die seemed compressed to a dot. I felt connected, not
only to the stars, but to all of nature and the cosmos. I felt a merging with
something much bigger than myself. Some kind of grand unity. After a
while I sat up and started the engine again. I didn't really know how long
I'd been lying there.

I've worked as a physicist for many years, and I've always held a purely
scientific view of the world. By that I mean that the world is completely
material, that it obeys forces and laws, and that everything in the physical world, including humans and stars, will eventually disintegrate and
reduce to their component parts. Even at the age of twelve or thirteen,
I was impressed by the logic and the materiality of the world. I had my
own homemade laboratory. One of the projects I built was a pendulum,
by attaching a fishing weight to a string. I'd read in *Popular Science* magazine or some similar magazine that the time for the pendulum to make
one complete swing was proportional to the square root of the length of
the string. With a stopwatch and a ruler, I verified this magnificent law.
Logic and pattern, cause and effect, as far as I could tell everything was
subject to analysis and quantitative tests. I saw no reason to believe in a
supernatural being, and that's still pretty much my view.

Yet after my experience in the boat in Maine many years later, I understood the powerful allure of the spiritual world. I understood the allure
of the non-material and the ethereal—things that are all encompassing,
unchanging, eternal, sacred. At the same time, perhaps paradoxically, I
remain a scientist. I remained committed to the material world. I have a
good friend in Cambodia who's a venerated Buddhist monk, and sometimes I talk to him about spiritual and scientific matters. The Buddhists,
as some of you may know, believe in the Four Noble Truths, and they

all deal with the cause of suffering and how to relieve that suffering. My monk friend made it clear to me that Buddhists come to the Four Noble Truths from their direct experience with the world, but there are other things that the Buddhists believe in, such as that the universe has gone through an infinite number of cycles, where their convictions are based exclusively on the words of the Buddha, a human being born with the name of Siddhartha Gautama, also known as the *lokavida*, the knower of worlds. I thought to myself, how do we know that the Buddha was the knower of worlds? Were Einstein and Darwin also knowers of worlds? The things that we believe about the physical universe and the spiritual universe, how do we come to that knowledge, and on what authority?

Science and religion differ profoundly in the ways the truths and beliefs are arrived at. In religion and theology, I think there are two origins for the truths and beliefs. First are the sacred books like the Bible and the Quran and the Vedas and the Pāli Canon. Believers have faith that these books embody the true Word of God or other enlightened beings, and if so, the authority of the teachings in those books comes from those beings. The second origin of religious truth and knowledge is more personal, what one might call the transcendent experience. So let me talk briefly about these two kinds of knowledge in religion, and then I'll talk about the kinds of knowledge in science.

Quotations from the sacred books are used to declare truths ranging from the origin of the universe to free will to the details of reproductive biology. Still today many religious thinkers attribute absolute authority and truth to the sacred books called divine revelation. Here's part of the pronouncement of the Second Vatican Council on Divine Revelation endorsed by Pope Paul VI, and I quote, "The books of scripture must be acknowledged as teaching solidly, faithfully and without error, that truth, which God wanted put into sacred writings for the sake of salvation."

I respect the notions of God and divine beings. However, I insist on one thing. I insist that statements made by such beings, including statements recorded in the sacred books, about the physical world have to be subject to test. In my view the truth of any statements about the physical world cannot be assumed. They must be tested and revised as needed.

The spiritual world has its own domain. The physical world is the province of science. In the physical world, laws cannot apply to some phenomena and not to others, or at some times and not at other times. It's not OK with me if the principles of aerodynamics work on some of my airplane flights and not others.

Most religions take God to have an existence outside of the physical world, beyond time and space. Because science is restricted to the physical world, it can never therefore disprove the existence of God. I am impatient with the so-called neoatheists like Richard Dawkins and Larry Krauss who challenge the belief in God by using scientific arguments. The neoatheists are missing the point in my view. Likewise, we can never prove the existence of God. Belief in God must be based on faith.

Let me turn to the transcendent experience. For me as a scientist and a humanist, the transcendent experience is the most powerful argument for a spiritual universe. By transcendent experience I mean the immediate and vital experience of being connected to something larger than ourselves, to feeling some unseen order in the universe. The experience I had looking up at the stars that night in Maine was a transcendent experience. I imagine that that many of you have had similar experiences. The transcendent experience may or may not involve a supreme being. Unlike the received wisdom from the sacred books or from the pronouncements of religious institutions, the transcendent experience is a very personal experience. And the authority of that experience rests in the experience itself. No other person can deny the validity of what you have felt. The feelings cannot be disproved. Most transcendent experiences are unique events that cannot be duplicated in the laboratory.

I want to now talk briefly about the kinds of knowledge in science and the way that we arrive at that knowledge. Scientific knowledge is of two types. One is the properties of physical objects, like the size and the mass of a raindrop. The other is the laws of nature.

One of the first human beings to formulate a law of nature was Archimedes, two thousand years ago. Here is his law of floating bodies, and I quote, "Any solid lighter than a fluid will, if placed in the fluid, be so far immersed that the weight of the solid will be equal to the weight of the fluid displaced."

We can speculate on how Archimedes arrived at his law. At the time, balance scales were available for weighing goods at the market, so Archimedes could have taken an object, weighed it on a balance scale, then put it in a rectangular pan of liquid, measured the height of the rise of the liquid, multiplied the area of the container by the rise in height and gotten the volume of liquid displaced. Then taken that same volume of liquid, put it in another container and weighed it. Undoubtedly, he would have done this experiment many times to arrive at his law, and he probably also did it with other liquids besides water, like mercury, to come to the generality of the law.

All laws of the physical world are like Archimedes law. They are precise, they are quantitative, and they apply to a large range of phenomena. Perhaps it's astonishing that nature should obey laws at all. On the other hand, it's possible that nature could not exist without laws, that there would be some fatal contradiction without laws—like two plus two equals four, and two plus two equals three. Certainly, an unlawful universe would be a frightening place to live in.

In the last two hundred years, we've discovered laws that govern everything from electricity and magnetism to the forces inside atoms to the expansion of the universe. From these laws we've been able to explain in quantitative detail everything from the orbits of planets to the color of the sky to the six-sided symmetry of snowflakes. We've seen no evidence to suggest that all of the phenomena in nature do not obey laws.

Here's a wonderful law of nature that you can verify for yourself when you get home. Take an object, drop it to the floor from a height of four feet and time how long it takes to hit the floor. You'll get about half a second. Then drop it from a height of eight feet. You'll get about seven tenths of a second. Drop it from a height of sixteen feet and you'll get about one second. After doing this a few more times you discover the rule that the time doubles for every quadrupling of the height. This was a rule found by Galileo in 1590, and with this rule you can now predict how long it will take any object to hit the floor when dropped from any height. You have discovered firsthand the lawfulness of the physical world.

The law of the swinging pendulum that I found when I was twelve or thirteen, I didn't believe it was true because I had read about it in a popular science magazine. I didn't believe it was true because I wanted it to be true or because the famous Galileo had said it was true. I believed it because it worked. It evidently represented some fundamental property of nature.

Let me sketch the methods and the manner of science. Galileo's law of falling bodies was found to be a special case of Newton's laws of motion and gravity. Newton discovered his law by analyzing the orbits of planets around the sun. For two centuries Newton's law worked beautifully, and then in the nineteenth century, very precise measurements with new telescopes showed that the orbit of the planet Mercury could not be completely explained in terms of Newton's law. There was a slight discrepancy. Very, very slight. Just one hundredth of one angular degree every century it was off.

However, Newton's law was so precise and the measurements were so careful that some scientists were worried. Then in 1915, Albert Einstein came up with a new theory of gravity that explained the orbit of Mercury and predicted many other phenomena, such as the existence of gravitational waves. Here's the main point. We know that even Einstein's theory of gravity, as subtle and as accurate as it is, will need revision.

All the laws of nature discovered by scientists are considered provisional. They're considered to be approximations to deeper laws. They are constantly being revised as new experimental data comes in, or there're new ideas that are tested. In fact, what we call laws of nature should really be called approximate laws of nature because even if there are final and perfect laws of nature, which many scientists believe, at any one time we know only approximations to those laws.

It's in the process of the testing and revision that we see the strongest differences between the methods and beliefs of science and those of religion. As I've said, everything that we know in the physical world, the province of science, is subject to revision. Everything must be tested and proved. The knowledge of religion—coming from either the sacred books or from the personal transcendent experience—is not subject to revision. It is not an approximation.

I want to end my comments by saying that there is something the scientists believe in that cannot be proved. I call it the central doctrine of science. The doctrine says that all of the phenomena of nature in the physical world are subject to laws, and those laws hold true everywhere and at every time. This doctrine must be accepted as a matter of faith, because no matter how logical and lawful the world has been up to now, we can't be certain that tomorrow there might not be a discovery or phenomenon that is fundamentally unlawful. Something that is fundamentally unexplainable.

When it was found that the orbit of Mercury did not quite follow the prediction of Newton's theory, scientists did not abandon their faith in the central doctrine. They did not attribute the discrepancy to some miracle or to the intervention of a whimsical God. Instead, they recognized a problem that required a more advanced understanding. In fact, I can't imagine any event in the physical universe that would cause scientists to label the event a miracle unexplainable by science. Even if Rebecca's chair suddenly started floating in the air right now, I would look for a magnetic levitator underneath it, or I would postulate some new force that was not yet known by science, but a lawful force, not a supernatural force. In sum, I would argue that despite the big differences in science and religion in the way that knowledge is obtained and revised, they both share a degree of faith, a belief and commitment to the unprovable.

Science experimentally tests and proves everything that it knows and believes about particular phenomena, but it cannot do so with the fundamental belief, the central doctrine. The central doctrine must be simply accepted.

I think science and religion are the two greatest forces that have shaped human civilization. Both of them reflect something deeply human in us. Certainly, they are both here to stay. So let the conversation continue.

R GOLDSTEIN: The vexed relationship between science and religion— the question of whether the differences between them are, in the language of a divorce court, irreconcilable differences—are issues that I think about pretty incessantly, going almost as far back as my memory reaches.

I was raised religiously and, unlike many others who identify as non-believers, I know deep down what it's like to believe with perfect faith in a God who not only created the grandeur of the universe, the beauty and harmony of its natural laws, but also a God who created each and every one of us in the grandeur of our individuality, giving us the capacity for good and evil. I know what it's like to believe in a God who has intentions regarding us, his creations, as to how we ought to live, who cares about how we live, and to believe that these divine intentions provide the necessary moral foundations for our lives.

But I long ago lost my faith in such a God. Even more deeply, I lost my faith in faith, which means that I don't miss it. Faith is not a mental effort I aim for. Quite the contrary. And I don't in any way feel that my life has any less meaning or purpose or has been deprived of moral obligations.

But I won't lie to you and say that the transition away from religion wasn't difficult. It severed me from a tradition that deeply connected me to people I love. I didn't want to believe differently from them, to live my life in ways that they wouldn't approve of. So I kept up the traditions and rituals far longer than my beliefs warranted. In some sense, this wasn't difficult, because the traditions and rituals were so deeply ingrained in me that for many years after I'd lost my faith, lost my faith in faith, the first thing I'd do when I got up in the morning, before getting my children off to school and rushing off to my job as a professor of philosophy, there to instruct my students in the rigors of self-critical analytic thinking, would be to automatically recite the Hebrew prayer that I had been taught to recite as a child of three, thanking God for restoring my soul to me. I had been taught that sleep is a kind of death, and so it's only through the considered intention of God that, day after day, one awakens to life. This was a bit terrifying, thinking that I kind of died every night. It might have had something to do with the sleep problems I had. But still it was a powerful message of the way in which my own little life was enmeshed with the will of Almighty God.

The experiences and reflections that pried me away from my faith in faith were of two kinds. I'll save the second kind for the end of the talk. The first had to do with my love of scientific explanation. This came from books. We were strict Sabbath observers, which meant we weren't allowed to do much of anything on the Sabbath. What we did was read. Books were expensive, and so we didn't own too many, certainly not children's books. But every Friday before the Sabbath began at sundown, my father took us to the library to get our Sabbath reading. And one week I picked up a book called *Our Friend the Atom*, and it blew me away.

I learned from this book that the world outside of my experience was very different from the way that it appeared inside my head. The world out there was composed of all these invisible little bits, my friends the atoms, which were themselves composed of little bits—electrons and positrons and neutrons—everything moving around in orderly ways, described by laws of nature that scientists had discovered, so that even solid-seeming things, like tables, had lots of empty spaces in them and were the scene of invisible constant motion.

The thought that the world out there was nothing like how we experienced it in the intimacy of our own minds was so astounding and thrilling it made me want to jump out of my skin. And as I got older and worked my way through that public library, I learned more thrilling things about the laws of nature that yield us a glimpse of the world out there so different from the way it seems. A high point was when I read Einstein's popular book on relativity theory, which corrects our intuitive ideas about space and time. My admiration for the human enterprise of science began to take the place of the awe I used to feel for religion's many claims about God's intentions.

It was natural for me to compare science and religion, since they overlap in certain ways. Both religion and science have views about the unseen aspects of the universe, of the forces that lie beyond what we can directly observe. In other words, both science and religion make claims about ontology, about what exists. Science has its ontology— as of this date, at the very least, it includes bosons and fermions, the four-dimensional manifold of space-time. And religion, too, has its ontology—certainly including a transcendent God, existing outside of the spatio-temporal world we inhabit, but with intentions focused on the spatio-temporal world, especially one part of it: us.

So both science and religion make claims about what the world beyond our experience is like. Science's ontology of the unseen is dictated by its formulating laws of nature to explain what we observe. Its underlying presumption is, as Alan explained, that nature is lawlike. Without this presumption there can be no science, just as Alan explained. Only rather than borrowing the language of religion and speaking of the necessary faith to which science commits us, I would describe our belief in the lawfulness of nature as a working hypothesis. It's the fundamental working hypothesis of science, because we have to presume it in order to do science. If we observe something that we can't explain in terms of the formulated laws of nature, then we go back to the drawing board and try to reformulate the laws of nature so that they are inclusive enough to account for all the phenomena we've observed. That's how science makes progress, by closer and closer approximations to the truth, always putting its grasp of the laws of nature to the test, by way of experiments, which is what makes science an empirical enterprise.

And this is something quite unique about science. Science, by its empirical methodology, enlists reality—the world outside our heads—as its collaborator. We set up experiments very carefully to give reality the opportunity to correct us. With experiments, we're going out on a limb, making our beliefs vulnerable to rebukes from reality. Oh, so you think that simultaneity is absolute, do you, that whether two events are simultaneous or not doesn't depend on which frame of reference you measure them in? Well, we'll just see about that! And voila—Einstein's relativity corrects us. "Well, we'll just see about that" can be taken as the motto of science.

But the presumption of nature's lawfulness is itself never put to the test. It can't be put to the test, since the whole procedure of putting theoretical propositions to the test presumes nature's lawfulness. We can't even grasp what it means to put a proposition to the test in the absence of taking nature's lawfulness for granted. But that doesn't make our belief in nature's lawfulness an article of faith, akin to religious faith. Rather, it makes it a scientific working hypothesis of the most fundamental kind. The whole enterprise of science is put to the test by our seeing how far this presumption of lawfulness gets us.

And as a working hypothesis, the presumption of nature's lawfulness has succeeded magnificently. Assuming nature's lawfulness has gotten us very far. We know more about the nature of the universe today than we used to know yesterday, and we'll know more about it tomorrow than we know today. And applying what we scientifically know—which is what technology is—has transformed our lives, for both better and worse. And the worst can be pretty bad. So, for example, there is the nuclear arsenal that we arguably never should have accumulated, or the personal computers with which we communicate with one another in social media formats that can bring out the worst in us. But the worst doesn't entail that science's way of doing ontology is inherently unreliable. In fact, quite the contrary. The mere fact that our knowledge of the laws of nature yields us technological applications that work, even if it doesn't necessarily grant us the wisdom to know how best to use the technology, provides retroactive confirmation on the working hypothesis of science.

In enlisting reality as a collaborator, science works into its very methodology a recognition of how deeply wrong our ontological intuitions can be. It has demonstrated how deeply wrong our intuitions are, even about such fundamental categories as space and time and causality. Einsteinian space-time is nothing like the Euclidean space and absolute time of our immediate experience. And causality in quantum mechanics is nothing like the relations of cause and effect—where both cause and effect must be local in relation to each other—that we take for granted. As I'd first begun to see from that wonderful book *Our Friend the Atom*, the world that the cumulative progress of science reveals to us is so radically different from the one we have in our heads that it can be very difficult to even wrap our heads around it.

The general lesson to be carried away from this is that our ontological intuitions aren't to be trusted, that we are sorely in need of a shared procedure that can correct them, a methodology that enlists reality itself as our collaborator. It is only science that does this.

So when it comes to ontology, science alone offers us a powerful reason to trust in it—which doesn't mean trusting that everything it says now is the final word, because that, in itself, would be to misunderstand the trustworthiness of the scientific enterprise.

But what about morality? Doesn't religion have it over science in this domain at least? Science might be superior in telling us what is. But we don't just want to know what is. Even more urgently, we want to know what matters. That's where our deepest human concerns are. We want to know, for example, whether truth matters, or whether rather it's power that matters, or whether rather it's virtue and justice that matter. Even more deeply, we want to know whether we ourselves matter, and if we do, what's the source of our mattering? Do we matter because we matter to God, are made in his image, or simply in virtue of being human, and if so, why? Is it our reasoning capacity that makes us matter, our capacity for moral judgments, our self-consciousness, or mere sentience? Do some of us matter more than others, or is mattering equitably distributed?

And science, which might well have the superior advantage in telling us what is, has almost nothing intelligent to say about what matters. True, the commitment to truth's mattering is implicit in the scientific enterprise, but science can't itself intelligently address this subject, can't validate its commitment to the truth's mattering, since it can't enlist reality as its collaborator in getting some evidence one way or the other. The very methodology that makes science superior in determining the facts of what is silences it in addressing what matters.

But does that mean there is simply no fact of the matter about what matters, including whether truth matters, no difference between one opinion and the next, so that you might as well go to an exorcist to get your poor troubled child some relief from her torments as to a child therapist? Or you might just as well be a malignant narcissist, intent on grabbing all the power you can, lying willy-nilly and recognizing no rights other than your own, as to be someone who cares deeply about the welfare of others and helps them as best you can? If you deny that there are any facts about what matters, because science itself can't establish these facts, then you really have no grounds for drawing such distinctions as, say, between the exorcist and the therapist, the malignant narcissist and the altruist.

The late paleontologist Steven Jay Gould had tried to make a place for religion by divvying up our opinions into those which involve facts and those which involve values, and he christened his bicameralism the non-overlapping magisteria, or NOMA. The magisterium of facts, he proclaimed, belongs to science, and the magisterium of values belongs to religion. He was recognizing that science can't in itself establish values, which is true, but his statement is nevertheless unfortunate, since it implies that there are no facts about values. There are no facts about what matters. But at the same time, by granting religion the authority over values, rather than other normative systems—say, humanism or hedonism, white supremacy or devil worship—he seems to be acknowledging that there are facts concerning what matters, which is why some normative systems are better than others. In other words, Gould's effort to accommodate both science and religion is incoherent, and it's incoherent because he doesn't take into account, in his tidy bicameralism, another self-correcting cognitive discipline, that does address itself specifically to facts about what matters, namely secular philosophy.

Lurking behind Gould's bicameralism is a philosophical view called positivism, which had first been formulated by philosophers, but then—in the self-correcting way that runs in parallel with science's own self-critical methodology—refuted by philosophers. Positivism is the claim that if a question can't be determined by empirical means, then there simply is no fact of the matter, and therefore any answer you endorse will be basically arbitrary. And positivism is what Gould's unfortunate statement presumes, in his having science usurp all facts, making his endorsement of religion over, say, Nazism, utterly arbitrary.

Rather than speaking of facts on the one hand and values on the other, which presumes positivism, it's better to speak about two different kinds of facts: facts concerning what is and facts concerning what matters. And our intuitions are quite faulty regarding both kinds of facts, with our faulty intuitions regarding what matters being far more serious in terms of how well our own lives and the lives of others fare. That's why it would be good if we had some kind of rigorous discipline that subjects our intuitions about what matters to the same self-critical scrutiny to which science subjects our intuitions about what is. And fortunately, there is such a discipline, and it's called philosophy. In particular, both epistemology and moral philosophy are directly focused on the facts of what matters. Gould's bicameralism between science and religion, itself grounded on a philosophically discredited positivism, entirely leaves out secular philosophy.

I had said there was another kind of experience I'd had which gradually led me away from any faith in faith, and I'll end with that. When I was a child, some distant relatives of ours escaped from Soviet-dominated Hungary and lived with us for a while. They brought with them the most amazing tales, first about how fairy-tale fantastic their lives once were, glittering with parties and picnics and trips down the Danube, and then, in a sudden reversal that was like something straight out of a child's worst nightmares, of how there were people everywhere trying to kill them, and how they hid first in this place and then in that place.

Of course, what I was hearing about, in complete incomprehension, was the Holocaust. I heard a lot of names that were very familiar to me, including my own name and those of my siblings and my cousins, and slowly I caught on to the fact that all of us kids had been named after people that hadn't found good hiding places.

That's when I realized how wrong people could be about what matters, that they could think that what matters is murdering people who belong to the wrong group. And I wondered whether there might be some enterprise, akin to the scientific, that could help us get clearer on the facts about what matters and who matters, that could demonstrate how wrong about the facts those people were who had come to the conclusion that all those people in my family that I would never meet simply didn't matter. And there is such an enterprise, requiring training, just as science does, that has helped us correct our faulty intuitions about what matters, demonstrating, for example, why truth matters, why justice matters, and which also offers corrections of our faulty intuitions about who matters—which happens to be all of us.

Religion can make us feel that we matter because we are living as we believe our version of God wants us to live. Or even if we aren't—if our will is weak and we're sinners, displeasing to our God—still we feel that we matter to God, even in his displeasure, and so, even in our sinfulness, we matter. And that's a powerful psychological feeling, but it's not the kind that subjects itself to any rigorous scrutiny, that offers itself up for correction, as all of our intuitions must, our ontological intuitions as well as our moral intuitions. In fact, our moral intuitions most essentially of all, because of weighty consequences.

M GLEISER: I'd like to ask you a couple of questions. In the play *Arcadia* by Tom Stoppard, he said something very meaningful, that goes beyond this idea of religion as being something that makes us matter. He says it is wanting to know that makes us matter, and I think that's wonderful because it's really about this inherent desire we have to know, secular or not, that makes us transcend the everyday sameness of life.

R GOLDSTEIN: We're the species that wants to know whether we matter. Wanting, even needing, to justify our own existence is distinctive about us. Is this very will to matter itself the thing that makes us matter? Or does it just make us some very presumptuous primates? Our will to matter motivates us to assert ourselves in all kinds of ways, some wonderful—art, science, mathematics, service to others—and some horrible—racism, sexism, classism, nationalism. If what [what] Stoppard means is that those who engage in the first kind of activities, asserting their will to matter in creative ways that allow them to transcend the everyday sameness of life, that [this is] what make us matter, then I worry that [this] leads to drawing a distinction between those of us who matter more and those of us who matter less. The secular answer to this question of human mattering has to have the universality of the religious answer—we all matter because God intentionally created each of us—while leaving God out of the picture. That's one of the things that gives religion such power, that its answer to the question of why we matter covers all of us. And it does so in such a simple manner. You don't have to go through any fancy philosophical reasoning to get to the conclusion of universal mattering. Philosophical reasoning yields that same conclusion—Spinoza, for one, proves it—but it's more taxing to the brain. So I certainly wasn't saying that religion alone can make us all feel that we matter.

But it is a powerful psychological source of the feeling of mattering, this feeling of mattering to God. It yields a person the sense of cosmically mattering, and what could be more gratifying? Religion packs a psychological whallop that makes it very powerful, which of course doesn't make it true.

M GLEISER: Right. But certainly meaningful to a lot of people. I want to throw a word at you both, because it is a word that I think a lot about, and that I think could be rescued from a strict religious context. It's been sort of hijacked by religion, and every time a scientist uses the s-word people go, "Really?" The word is, you guessed it, spirituality. I think spirituality is a wonderful word because really, if you think about it, it comes from inspire, aspire. It's about inhaling the world into yourself to expand who you are. Spirits and spirituality are completely different things, at least to me anyway, and I just wanted to know what your take on this is, and the possibility that science and spirituality are much closer to one another than people may think. Alan talks a lot about this in his book.

A LIGHTMAN: Well, first of all, I think that spirituality may or may not involve an all-powerful creator. For me spirituality is very connected with the transcendent experience. It's the feeling of being connected to something larger than ourselves. That there are eternal values, there are things that we believe in. For me, beauty is part of the spiritual world. I don't see any contradiction between the notion of spirituality and the enterprise of science.

M GLEISER: So that experience you had in that boat, would you call that a spiritual experience?

A LIGHTMAN: Yes, I would.

M GLEISER: OK. So what about you Rebecca? Now I put you on the spot. Have you ever had something like this?

R GOLDSTEIN: It's a squishy word, spirituality. If it makes a claim, an ontological claim, as to the existence of spirits, then I'm not interested. But if it means simply a certain psychological experience of being drawn out of yourself into something far grander, so that you lose all sense of your own particularity, lost in wondrous admiration for something else—whether a beautiful piece of music or beautiful mathematical proof or Einstein's theory of general relativity or the starry night—then yes, of course, this is a psychological experience and a deeply beautiful one, and yes, I'm rather prone toward it. In that sense, not only is science consistent with spirituality, but it can be the very source. And it's perfectly compatible with being an atheist, too.

Spinoza is very relevant here, because Spinoza, a seventeenth-century philosopher who was excommunicated by his Jewish community [in] Amsterdam, and who therefore is my favorite philosopher, was accused of being an atheist but also of being God-intoxicated. The poet Novalis called him God-intoxicated. The reason for such confusion is that he identified God with the natural order itself. He thinks this natural order, the totality of natural laws, ultimately explains itself. We can't actually see how it explains itself, because we can't grasp its totality, but nevertheless we can know that it does. He has a proof for this, which is subtle and itself quite beautiful. But the emotional impact, he says, of realizing that we live in such a world, in which the laws of nature are complete unto themselves, so complete as to determine their own realization, is the kind of transcendent experience we're talking about. And Einstein whenever he was asked about whether he believed in God, he would say, "I believe in Spinoza's God," which is a very tricky way of saying, "I don't believe in a God outside of the universe, who created it. I believe only in the natural universe. But it's a natural universe which, in the beauty and harmony of its laws, is worthy of awe." That's a kind of spirituality.

A LIGHTMAN: Well also, Einstein had a very modest view of our understanding of God. He said that any thinking person should not rule out the possibility of God. He said that we are like children coming into a vast library and seeing all of these books, and not understanding them but knowing that something wrote those books. Some intelligence wrote those books. That is what he said when he was asked about his view of God.

M GLEISER: That's his belief in a rational, lawful universe.

R GOLDSTEIN: Yes, I agree. It's his belief in a rational, lawful universe. He said the idea of an anthropomorphic God who actually cares about us is childish.

A LIGHTMAN: But it also suggests—

R GOLDSTEIN: It's Spinoza's God he's talking about.

A LIGHTMAN: But I think his statement also suggests that there are mysteries that are beyond our grasp, so that's not a contradiction with the statement that it's a lawful universe.

R GOLDSTEIN: Absolutely. Again, this is Spinoza's view, that there is a final theory of everything, all of the laws, perhaps quite small in number, that explain everything that exists, but it would take a mind more powerful than ours to be able to know it. We get just enough of a glimpse to understand all that we're missing. So that complete theory exists, theoretically speaking, but practically speaking, we can never grasp it ourselves.

A LIGHTMAN: Well, we would never know whether we had the final theory.

R GOLDSTEIN: We would. If we had it, we would know we had it, because we would be able to prove that such a set of laws would entail its own realization. It's what Stephen Hawking was speaking about at the end of *A Brief History of Time*, when he wrote that Einstein was saying we didn't yet have what was the only possible universe. All others would be ruled out. It would explain its own existence. That, at least, is Spinoza's view, which—

A LIGHTMAN: But you could never be sure that there wouldn't be some phenomena that you would observe tomorrow that would be unexplainable with your theory that you—

M GLEISER: That's the scientist speaking, but I think the philosopher is talking about this something, which is beyond. I wrote a whole book called *The Island of Knowledge*, which is exactly about this. That the fact that there is no way you can actually get to a final theory of everything because you'll never know where you are in the big scheme of things.

R GOLDSTEIN: But, if the Spinoza-Einstein-Hawking idea is true, then we can be assured, even if we'll never grasp for ourselves the set of laws that entails its own realization, we can be assured still that it exists.

M GLEISER: Isn't that what's so frustrating about the human condition? That we have this spark of divinity, but it's just a spark. We can't really grasp the whole thing. There are these glimpses of going deep into something, yes, you know, but is it everything? It could be. Smells like it, and then BAM, no. There's a new particle. Everything is ruined. Got to start again. I think that's the—

R GOLDSTEIN: If we really had it.

M GLEISER: Then we'd know.

R GOLDSTEIN: Then we would know. There would be nothing arbitrary about it. That's the sign one is looking for in the final theory—the elimination of all arbitrary elements.

M GLEISER: Right. Exactly.

R GOLDSTEIN: You're right. Our science is filled with things that are arbitrary. It's that way because it's that way, but we don't know why.

M GLEISER: I call it monotheistic science.

R GOLDSTEIN: You call it what?

M GLEISER: Monotheistic science—to get to the theory of everything. Steven Weinberg, this wonderful physicist, wrote a book called *Dreams of a Final Theory*, where, at the time, he truly believed we were getting there. In fact, there are lots of our colleagues [who] believe that. A few years back we were almost there, we thought: four fundamental forces, superstrings, etc., we're going to do it. And you just can't do it. You cannot. Even if you do get to understand superstrings and all this beautiful stuff, you cannot know that the day after tomorrow—

R GOLDSTEIN: We know that we don't know.

M GLEISER: Well, not a lot of people.

R GOLDSTEIN: It's not amazing that we don't know everything. It's amazing that we know anything; that we know as much as we know.

A LIGHTMAN: Einstein also said that the most amazing thing about the universe is that it's comprehensible, right?

R GOLDSTEIN: That's extraordinary. We're evolved apes, whose primitive cognitive capacities, evolved to help us survive, have managed to take off into the kind of abstract domains we need to grasp to understand something of the universe. It's a kind of humility to realize that the intuitions that we bring to bear are not infallible and our cognitive capacities aren't guaranteed to see their way to the end of the story, the final theory.

I don't like using the word faith when it comes to our belief that nature is nomological, that is, that it's lawlike. It's true that we can't prove it. We can't prove that nature will continue to be governed by the laws of nature, although of course, if that happens, we won't survive to witness the nomological breakdown, since the laws of biology, of neurophysiology, are essential for us being us. But we can't prove that in the future there won't be such laws. In fact, we can prove that we can't prove it. That's what David Hume, the great eighteenth century Scottish philosopher, demonstrated. He's the one who proved conclusively that there's no way to justify what he called the uniformity of nature—that the laws of nature won't simply change. There's no way to do it because we have two ways of justifying propositions, through either pure deductive logic, or through induction. Since the negation of the uniformity of nature doesn't entail a contradiction, that means it [untrue] by reason of pure logic. So maybe we can prove it inductively. But you see, that involves one in circular reasoning, because the very practice of induction presupposes that what has been uniformly true in the past will continue to be true in the future—the very thing we're trying to prove. So there's no way to justify induction. But then, too, there's no way to non-circularly justify deductive logic either, since one has to use logic in order to prove anything. So this is a lesson for us, long the subject matter of philosophy, that the very practices that define reason itself, deductive and inductive, can't justify themselves, just because they define reason itself. There's simply no thinking without them. But this is a very different situation from the kind of thing that one means in speaking of faith. In the one situation, the thought practices are so basic that there's no coherent thinking without them, and in the other case, well you can think quite coherently without faith.

A LIGHTMAN: I don't agree with that. I mean it could be that the world is 99.99 percent logical, and it's logical enough and lawful enough for us to exist. In a similar way, there are some interactions with elementary particles in which a certain symmetry is violated only once out of many billion interactions. It could be that all of the biology that produces us lives happily in that 99.99 percent of the world that is lawful.

R GOLDSTEIN: Some laws are probabilistic. Not all laws have to be absolute, so that there are, as you know, probabilistic laws. And it's in the nature of probabilistic laws to sometimes be violated. That's in the nature of probabilistic laws, but that's not a violation of the law.

A LIGHTMAN: No. No.

R GOLDSTEIN: We probably don't disagree here, except on whether we can use this same word, faith, for the situation of faith in the religious ontology, no matter how much comfort and meaning it might yield to some people's life, and the epistemological situation we are in as regards the basics of thinking, without which we can't even make sense of the difference between coherence and incoherence. I feel that when we use the word faith for that sort of thing, this condition of doing science, but of just making sense at all, I mean you know, pursuing our lives. To use that same word suggests a false equivalence between a kind of faith that is part of just pursuing rationality, and faith as it's usually used which is—

M GLEISER: Religious.

R GOLDSTEIN: I want the world to be this way very, very much, and therefore it is.

M GLEISER: Right. Yes. Exactly. On that note, and this is pertinent, this question. It's a question for Alan but it really is for all of us. You mentioned that a universe without laws would be a frightening place, but a universe with laws like ours can also be frightening. Look at the hurricanes. Are there any scientific truths or future scientific inquiries that threaten your sensibilities as a humanist?

A LIGHTMAN: Is that a question?

M GLEISER: Yes. A question. Are there any scientific truths or fields of scientific inquiry that threaten your sensibilities as a humanist? I know one, but I'll be quiet for now.

A LIGHTMAN: I can't think of any. I'm sorry. I'm not able to come up with it.

M GLEISER: What about—

R GOLDSTEIN: What about entropy? That's a scientific truth that's going to get us all in the end.

M GLEISER: No. No. I mean, I was thinking more like in terms of fields, but yeah entropy sucks, really. I was thinking more in terms of fields of scientific inquiry. I was thinking about transhumanism.

A LIGHTMAN: Well, there [are] plenty of fields of scientific inquiry that have emotional resonance with me. Just the neuroscience and the exploration of what thinking is. The question of the big bang and whether something came before the big bang. These are certainly issues that affect me in my gut, but I don't feel threatened. Maybe I am putting too much emphasis on your word threatened.

R GOLDSTEIN: What do you think?

M GLEISER: I was thinking of transhumanism—transhumanism being the idea that we are in the process of becoming something else, a different kind of species altogether—because of our coexistence and symbiotic relationship with digital technology. For example, picture going to work one day and you're stuck in traffic, and suddenly you realize that you forgot your cell phone at home. You can't go back because otherwise you'll be late to work, so you have to spend a whole day without a cell phone. It's desperation for most people, because the cell phone is part of you. It has become an extension of your person; you're not yourself without it anymore because you need it so badly. In fact, if you look at people's different cell phones, you're going to see that there will be lots of apps which are the same, but some of them are going to be very particular to you and to your tastes as a person, almost like a fingerprint. Smart phones really are a digital extension of you. So you are both human and machine already.

The transhumanism movement goes way beyond that, as I'm sure you've seen in sci-fi movies, where we actually become machines, in a sense that we will expand our neural capacities and physical strength, possibly to a point that we will be almost unrecognizable as human. We will be something different.

A LIGHTMAN: I see. OK. Well, yes that does—

R GOLDSTEIN: I think you're talking about technology, the application of scientific truths, rather than scientific truths. And there are lots of applications that could make us uneasy as humanists.

M GLEISER: Yes. Well, fields of scientific inquiry.

R GOLDSTEIN: Designer babies, right? Using CRISPR to design the genetic makeup of our offspring. I think that's really frightening.

M GLEISER: That would be a nice next dialogue honestly. Designer babies, what a great line.

R GOLDSTEIN: Designer babies? Yes, that's a good line. I didn't think it up though.

A LIGHTMAN: I think the smartphones that we have right now threaten me as a human being because everybody's plugged in.

M GLEISER: Alan is famously resistant to email by the way. He's resisting the onslaught, which is very brave.

There is one question here that says, "Do you think that a transcendent experience could come to someone who had never heard of God's Spirit or religion, and if so how might that experience differ from someone raised in a traditional Western society?" So is there some sort of universality to the transcendent experience?

A LIGHTMAN: I think definitely.

R GOLDSTEIN: Once I was at the Cape of Good Hope, staring out into the ocean at a magnificent sight, a perfect rainbow formed right out there on the ocean. I was staring transfixed, having a deeply transcendent experience, and then I noticed right in front of me, there was a line of baboons and they were all facing out at the rainbow emerging out of the sea and as deeply transfixed as I was. Baboons are very obstreperous creatures, always moving and making trouble, but they'd been reduced to complete silence. When I realized that I'd been sharing that transcendent experience with a troop of baboons, well that was yet another order of transcendence.

A LIGHTMAN: But do they believe in God?

M GLEISER: Maybe the rainbow was God for them. Next question: "Given recent theories regarding the nature of matter, is there speculation that individuals may contain information in the form of memory of lives and events that occurred in previous eras? That is to say, could there be an energetic form of 'DNA' that is held in the relational elements of matter or energy in the body, a sort of atomic memory, so to speak?"

Lots of Tibetan Buddhists believe in things like that, and so this would be like a theory of reincarnation of some sort. I don't know what you think of all this, both of you.

R GOLDSTEIN: I have no experience with these sorts of things. I know that it's reported sometimes, these anomalous events, children born seeming to remember things, and if scientists became convinced that this was really a phenomenon, that there was no other way of explaining what these children seem to know, and, you know, no hanky panky in the reports, well then we'd have to start looking for the laws of nature that could accommodate the phenomena, and what the questioner mentioned, not that I could follow it, but maybe it could be part of the answer or even the answer. But no, I really have nothing intelligent to say, as I've just demonstrated.

M GLEISER: I have one more final question or idea to discuss, which is this notion of supernatural versus natural. I want to know what you think of this. I'd argue that supernatural things don't really make a lot of sense for the following reason. Let's imagine that I'm in my bathroom and I see a ghost. The fact that I saw a ghost means that whatever the entity is, it radiated some form of electromagnetic energy that impacted my eyes and impressed my visual cortex so that I could see it. In this sense, it seems to be a perfectly natural phenomenon, even if mysterious. To have an experience [that] is supernatural makes it natural, because it involves an exchange of information, which in principle supernatural things should not be able to do, because they inhabit a different kind of reality.

A LIGHTMAN: What if you saw the ghost and it was not through electromagnetic radiation? See, that's a term of science. So what if somehow you were able to see the ghost without . . . even with your eyes closed?

M GLEISER: So I saw it in my mind, so to speak. That'll be different. If that's possible then I would credit it to some sort of hallucination. As it would be coming from my head, you know?

A LIGHTMAN: Because you're committed to the central doctrine of science.

M GLEISER: I'm hearing voices. You know it's coming from an outside source. If not, I'm probably in trouble and need to see a psychiatrist. Likewise, if I'm seeing things inside my head, something must be wrong. Do you agree? The idea of a supernatural reality parallel to ours is a complicated one from the physical material perspective.

A LIGHTMAN: Well, but to know what supernatural is, you have to first know what natural is.

M GLEISER: Very true. And that's why we shouldn't simply discard the possibility with an arrogant rational dismissal, although we must keep our wits sharp at all times. I mean, where is the boundary between the two?

R GOLDSTEIN: So this is not to exclude the possibility of the supernatural existing, but just the grounds that you could possibly have for thinking that it exists. For any individual, it's much more likely that one is having an unfortunate psychotic experience. But what if we all saw it, all saw the same thing, agreed afterward it was the same thing. Is there anything really incomprehensible here? I don't think there's anything incomprehensible about this idea. Suppose one wouldn't even have to change the laws of nature. It's just that the boundary conditions that would produce the phenomenon would be so special and unlikely. I guess if my chair levitated, I too would not—

M GLEISER: Believe in the supernatural.

R GOLDSTEIN: I would not but if—

M GLEISER: You'd be scared.

R GOLDSTEIN: I'd be holding on. I mean, if we suddenly walked outside and looked up at the sky and the stars were regrouped into a big Jewish star, a Magen David, well that would give us something to think about.

M GLEISER: But we'd better wait sitting down for that one.

R GOLDSTEIN: Yes.

M GLEISER: But those experiences, the one you, Alan, had in the boat, they are not . . . they're very natural. I mean, these feelings of transcendence are perfectly natural. I remember I had certainly one, maybe more than one experience like that when I was fifteen. My older brother Luiz was there. We went to spend the weekend together at these spectacular islands south of Rio, where I grew up. At the time, I was this very mystical boy, with a foot in reality and another in fantasy. He said, "You want mystical experiences? Sit on that rock in front of the ocean and watch in silence." I was there for a very long time, and I could feel precisely this increasing sense of material dissociation, of my individual essence disappearing to become a feeling of oneness. It was a truly moving experience for me. Those things happen but they have nothing to do with a supernatural dimension, I think, even if they invite you to dive deeper into the meaning of reality. I guess I took the plunge then, and am still searching.

5

THE MYSTERY OF TIME

A Dialogue Between a Science Historian and a Physicist

MGLEISER: This conversation is on a subject that is profoundly commonplace and profoundly mysterious: time.

Everybody knows about time. We all talk about time, complain that we never have enough of it, and puzzle about its nature. One thing I observe is that the older you get, the faster time seems to pass—a great injustice if you ask me. Is this really true? Does this feeling of time's elusiveness have any relation to the physicist's definition of time? Or is it something completely different and subjective? Could there be two kinds of time, one being the precise time of science and the other a more personal and psychological time? And if there are, can they be made compatible with each other? Or is subjective time not really amenable to a scientific description? And here is an even harder question (there are many!): Does a scientific description of time help us understand the nature of time at all? Or does it remain mysterious and ultimately subjective, even if we measure it ever more precisely?

Jimena Canales grew up in Mexico, then came to the United States, where she did a master's and a PhD at Harvard in history of science. She stayed at Harvard, teaching for many years, and then she moved to the University of Illinois as the Thomas Siebel Professor of the History of Science. She recently returned to Cambridge, Massachusetts, to dedicate herself to writing. She's the author of many technical articles and journals, and also of two books, *A Tenth of a Second: A History* and *The Physicist and the Philosopher*.

The Physicist and the Philosopher has received much praise. It was voted one of the best science books of 2015 by NPR's *Science Friday* and by Maria Popova's *Brain Pickings*. It was listed as a Top Read by the *Independent* and named one of the Books of the Year by the *Tablet*. Jimena's first book, *A Tenth of a Second*, was listed as one of the *Guardian*'s Top 10 Books About Time. She has also written many essays for the *New Yorker*, the *Atlantic*, *Wired*, the BBC, and other venues.

In *The Physicist and the Philosopher*, the physicist is Albert Einstein and the philosopher is Henri Bergson, who was the biggest intellectual superstar in Europe in the 1920s. Believe it or not, Bergson was much more famous than Einstein at the time. On April 6, 1922, the two sat together in conversation in Paris, to address the nature of time. What transpired that evening would change our understanding of what time is.

Professor Paul Davies is a world-renowned theoretical physicist. He works in cosmology, in astrobiology, and also in astrophysics and the theory of black holes. He has written many books, and he's also a Regents Professor here at ASU and the director of the Beyond Center for Fundamental Concepts in Science.

He has written three books explicitly on time. One of them, a more technical one from 1974, is called *The Physics of Time Asymmetry*. And then two others addressed at a more general audience, *About Time* from 1995, and, the one that I would love to know the answer to, *How to Build a Time Machine* from 2001. Paul may not even be here right now—it may be his future self, visiting us.

He's won many awards, including the Templeton Prize, and also the [William Thomson, Lord] Kelvin Medal and the Faraday Prize, which are given to people who have a true impact on the public understanding of science.

J CANALES: I'm a historian of science. I've been studying science for about twenty years, mainly from a historical perspective, although I'm very interested in contemporary science. I started my career writing my first book, which is *A Tenth of a Second: A History*, and I joke that I wrote the longest book about the shortest time period in history. It's about why we started to be concerned with measuring short periods of time that don't really matter for our everyday lives.

The book starts around the middle of the nineteenth century, when a French astronomer, François Arago, who claimed that clocks that measured the tenth of the second were no longer a "vain luxury," but that they should be adopted in science. After that we see a race to measure even shorter periods of time. And I believe we are still in that race.

From that time on, the ability to measure short periods of time and to have accurate standards of measurement became a mark of civilization and civilized cultures. That was the main interest in my book.

As a young historian when I started writing it, I read a lot of political history. I read about presidents and kings. And I thought that none of the history that I was reading, including most of the history that I was being taught in my classes, explained the essential texture of modernity. Why are our fingers now on keys, why are our eyes tied to screens? How did it come to be that we became surrounded by different systems in which we have a stimulus, which goes through our nerves, and to which we react within a tenth of a second? That's the importance of the number that I was interested in. These systems came to represent modernity, from telegraph keys to automatic pedals. Why did they become so prevalent?

I look at these time-keeping instruments, at the history of reaction time, at the history of astronomy, together with the history of experimental psychology. In my view, these interrelated disciplines could help us understand the texture of modernity, the sensory-motor order of modernity—why our eyes and fingers end up in certain places, manning certain devices and not others.

Toward the end of the book, from my research I knew that one of the most famous philosophers of the late nineteenth and early twentieth centuries was Henri Bergson, a French philosopher.

He was basically more famous than Einstein, and he was an expert on time, among other things. He wrote a book in 1907 called *Creative Evolution* that made him world-famous. And he was a real public intellectual. He was one of the first intellectuals to go on the radio. He hobnobbed with heads of state. He was influential in convincing President Wilson to join the war. He's rumored to have created one of the first traffic jams when he gave a lecture in New York City.

William James, for example, admired him. And I remember when I started studying Bergson, people would say, "Well, he read William James," and I would say, "No no, William James read Henri Bergson." So I knew of this figure [who] was largely unknown today. Except for one conference in Berkeley, probably about fifteen years ago, but there hadn't been a major conference on Henri Bergson in the last fifteen years in the United States.

And I bumped into this document where there was a transcript of a meeting where Einstein was giving a talk in Paris and Henri Bergson was in the audience. I just couldn't believe it. I thought this was a historical treasure, and I thought that there must have been many, many people who had written about this. We all know Einstein's importance, and some people knew of Henri Bergson's importance. So what did they say to each other, and what happened at the meeting?

And to my surprise, there wasn't any book-length account of their meeting on April 6, 1922. So I decided to dive in and to figure out as much as I could about that day, and to answer some of the questions I had. Why is Bergson so little-known today? It's really a fascinating story.

Einstein had become a very famous physicist in 1919, after an eclipse expedition confirmed his general theory of relativity. Basically, it confirmed that gravity bent light around the sun. Some British astronomers, Frank Dyson and Arthur Eddington, measured this, and they did one of the first announcements of science by press release—a big event. Their observations could have been explained in many ways. There were many other people who had been interested in the possibility of gravity bending light waves. But the way that the paper was presented, there were only three options about what the results could mean. In the first option, there was no change at all, gravity did not affect light waves. There was another option where it affected light waves to a certain degree, and there was a third option where it bent the light waves to the degree predicted by Einstein, and the results matched number three.

Newspapers all over the world covered the incident. And Einstein went from being a well-known, respected but controversial physicist to being a worldwide star. He was chased around by photographers. He wrote that at that time every child in Berlin could recognize him from photographs. His face was more recognizable than Hollywood stars at the time. This [was] the nineteen teens, the beginning of Hollywood, the beginning of the movie industry. There [was] an incident in which he [went] to California and [went] to the movies, and they stopped the movie because the biggest Hollywood star at the time, Mary Pickford, wanted to say hi to Einstein. And Einstein shook her hand and didn't know who she was.

So there's this very interesting reversal at the time of the role of the scientist as a public intellectual. Helmholtz, Pasteur, Darwin—they were very well-known to the public, but they weren't really covered regularly by newspapers. This is the time when the daily press became one of the venues in which people learned about science. It [was] after World War I, and readers [were] eager to hear wonderful stories, not just stories about carnage and devastation and the bad Germans or the bad French guys. Science started taking this role.

Einstein [was] invited to go to France to speak, particularly because he [was] a pacifist and had written against the remilitarization of Germany. He was a German Jew who worked in Switzerland. He had complicated relationships at home. He [was] very fond of the British astronomers who ran the eclipse expedition. There [were] people in France who really wanted to support his theories. Surprisingly, these [were] philosophers more than physicists.

Einstein [was] not invited by the Société de Physique. He [was] invited to give some talks at the Collège de France, and also—more importantly, the meeting that I'm interested in—at the Société Française de Philosophie on April 6, 1922. There are some photographs of his talks, of people trying to get in through the gates, trying to see this great star.

He [got] to Paris by train. He arrived in the City of Lights at the Gare du Nord, and there [were] many journalists waiting for him. So he escaped through the other tracks, bypassing where it says "Do Not Enter," to avoid the crowds.

This [was] a huge event. He started speaking at the Société Francaise de Philosophie in very bad French. He didn't speak French well at all. We all know this myth that Einstein was not good at school, and it's pretty much a myth except when it came to speaking French. He really did not like the language. One of the first references that I found of Einstein talking about Bergson was that a friend of his wanted to read something in French in order for them to get better at it. And Einstein [said], "No, I'm not going to read that flaccid Bergson."

But many years later, there he [was]. His correspondence also shows that he [was] very nervous about the talk. He [was] very nervous about speaking in French. He had a good friend, Paul Langevin, who developed what we know now as the twin paradox, and the journalists said that Paul Langevin was whispering the answers to Einstein that evening.

There is a complete transcript, and this was my excitement as a historian. On top of everything, here I have a transcript of everything they said! Sometimes even the tone of what they [said] is conveyed in parentheses in the transcript. The evening started well with a short introduction by Einstein and then they opened up the discussion.

And Einstein [said] something that served as a detonator that evening. Henri Bergson [was] in the audience, and they [were] talking about the nature of time and what is subjective time and Einstein said, "*Il n'y a donc pas temps de philosophes*" (There is no time of the philosophers). People started nudging each other and a member [told] the organizer, since "Bergson was among us" it would be nice for "Bergson himself to take the floor." Bergson [said], "I came here to listen. I did not come with the intention of speaking, but I will cede to the polite insistence of the Société de Philosophie." And he responded in about fifteen minutes to this incendiary claim that there is no time of the philosophers.

I'll say a little bit more about what was at stake that evening, and the response it received. But let me backtrack a little bit and talk about what Bergson found so offensive about this, and what happened afterward.

Bergson ended up writing a complete book, called *Duration and Simultaneity* against Einstein's theory of relativity. The book had already been in production; Bergson had studied Einstein's work for more than a decade. He had attended one of the lectures of Paul Langevin in Bologna that first discussed the twin paradox, and the book would appear a few months after their debate at the end of the year 1922.

That year also, Einstein received his Nobel Prize. He had been hoping to get the Nobel Prize for many years. He had even promised to give the money of the prize to his ex-wife for alimony. But when he received the prize, he didn't get it for the theory that had made him famous, the theory of relativity. He got it mainly for his work on the photoelectric effect, and also for his work on Brownian motion.

The presenter of the Nobel Prize was very explicit. He said that the reason why the prize was not given for relativity was because everybody knew that the philosopher Bergson had contested this theory in Paris, and that the verdict toward it being right or wrong was not an experimental verdict. It wasn't an experiment that needed to happen (That had happened with the eclipse expedition. If one was looking for an experimental verification, one could have said that [it] happened. The Michelson-Morley experiments had [also] been done decades before). The presenter said that the verdict had to be epistemological. They were looking for a solution that was not technical, in order to know if scientists should adopt, or not, the theory of relativity.

There's a lot that has been written about the experimental verification of Einstein's theory of relativity, and I absolutely do not contest it. But at the time, there were many scientists—the most important scientists to have worked on this topic, namely Henri Poincaré and Hendrik Lorentz—who had looked at the same experimental results and concluded, "Yes, this is true and what Einstein says is true, but we do not think that we need to completely revolutionize our everyday notions of time and space. We can explain the bending of light by gravity, we can explain time dilation, we can explain length dilation in the same ways that Einstein and the experimental physicists are explaining it. But let us not abandon the way that we think about space and time in our everyday lives, because that makes sense for us."

As many of you know, the way that Einstein explained space and time entails some real paradoxes. There are some issues that are very hard for us to wrap our heads around. Einstein did not care about those. He basically said that the way that we experience the flow of time is an illusion. When his friend Michele Besso died, he wrote a famous note to [Besso's] sister saying, "For the physicist of faith, the difference between the past, the present, and the future is nothing but a persistent illusion."

So it was due to our sensorial limitations, but if we thought as physicists—if we used mathematics—according to Einstein, we reached this other theory of space and time, in which space and time had four dimensions. They were put on equal footing, according to the general theory of relativity. That's obviously hard for us to understand. We can get lost and end up on the west side and not [the] east side, but you will never find yourself lost and find yourself in yesterday or tomorrow.

For us, there are some genuine differences between how we experience space and time, in addition to other paradoxes, particularly the twin paradox. I won't say a lot about that, but it's this famous example that you might remember from high school, in which you have two twins on Earth and [if] one of them leaves Earth and travels close to the speed of light, he will have aged less than the twin [who] stayed stuck on our planet. If they return, he will ascertain that.

Bergson, in his book, wrote about this paradox, and he said the twin that travels is a fictional twin; it's fiction. And that is wrong, according to science. So when Bergson's book *Duration and Simultaneity* appeared in 1922, Einstein wrote many letters saying that Bergson had made a mistake, and that his mistake had been a mistake of physics. That represented the moment in which you had the towering figure, Henri Bergson, who [was] a great public intellectual, basically pass off the baton for talking about the nature of time to a physicist.

Today if most of you want to learn about time, you might read St. Augustine, or you might read a poet, but if you want to figure out what it is, you [will] probably go and read Stephen Hawking or another physicist. I'm very interested in this change in authority in who can speak for time.

Just to wrap up, one of the things that I wanted to do in my book is that I didn't want to write another book about what time was, or about the nature of time. I thought that had been done a lot. I wanted to write a book about who has the authority to speak for time. And how do we as a community, as a society, grant that authority? And how is that authority sustained by certain professions, by certain personalities, and disseminated in a particular way?

P DAVIES: Time is an endlessly fascinating subject. Everybody thinks about it, and I've given so many talks on it I even took the trouble some years ago to buy a special tie for occasions like this, which I am wearing now. And people of course always want to know, could I go back in time and fix up those exam grades, or decide not to date that girl who seemed so nice at the time. And we use expressions like "And I will love you till the end of time." The end of time? What's that? Is there an end of time? Could there be a beginning of time? What about God? Outside of time or eternal? What does that mean? What does eternal life mean? Forever and ever and ever? We bandy these terms around, but I'm a physicist, and I want to bring some precision into this discussion.

I first got interested in the nature of time in a previous millennium. It was actually 1967. I went to a lecture at the Royal Society in London by Fred Hoyle, the British cosmologist. And what Fred said was, set up a radio station, transmit a program, and the listener gets to hear it a little bit later than the radio station transmits it, because the radio waves have to propagate across the space between. The waves never arrive a little bit earlier. But the laws that describe how radio waves move, actuated by them, are symmetric in time. They make no distinction between past and future.

According to the laws of physics, there's nothing wrong with listening to a radio program before it is transmitted, rather than after. And the question in Hoyle's mind was, how do we resolve what seems like a paradox? That in daily life we have this overwhelming impression of an arrow time, a directionality of events. The future seems different than the past in many ways. Pick a movie or any everyday scene, play it backward, everybody laughs, because it's so obviously preposterous. What is it that breaks that symmetry of the underlying laws and gives an arrow of time in the everyday world?

I want to come back to that because I'd like to actually start with a deeper question, and in many ways a tougher question, which is, did time have a beginning? Was there an origin of time? I just said the end of time, what about the beginning of time? Does it stretch back forever and ever in the past? I used to lie awake as a teenager worrying, "Well, if time has gone on forever, why am I living now, rather than some other time?"

These days we tend to think of the origin of time in connection with cosmology, with the origin of the universe, the big bang that started it all off. When I was a student (in a previous millennium remember) the party line was that the big bang was the origin, not just of matter and of energy, but of space and time as well. That is, there was no time before the big bang. The big bang was the beginning of time.

If I discuss this at dinner parties, people say, "Oh but that's a trick, something must've happened before the big bang. Or there must've been something going on prior to this. Time couldn't have begun there." But actually, it's an old idea that time had a beginning. Augustine, in the fifth century, already said that the world was made with time, not in time. In fact, he wasn't the first person to say that.

The idea that time itself comes into being with physical reality, with the universe, is a very ancient concept. According to the simplest version of modern cosmology, the big bang theory, that's what happened.

Now, I don't have time to get into it, but you could say, "Well, why did time just switch on like that at some arbitrary moment? What was it that caused that to happen? Can we make sense of that?" Well, I should just say in passing that Stephen Hawking and James Hartle developed a theory in the 1980s where time is bounded in the past. Paradoxically, however, although time doesn't go back forever and ever, there's no particular magic moment, no event, no singularity at which time and space abruptly switched on. The reason for that has to do with quantum physics. I can't get into that now, or you'd be into a tutorial, but I want to emphasize that in the simplest version of modern cosmology, the big bang was the origin of time, and asking what happened before the big bang is, as Stephen Hawking put it, "like asking what lies north of the North Pole?" The answer is nothing. Not because there's some mysterious Land of Nothing north of the North Pole, but because there [is] no such place as "north of the North Pole." For the same basic geometrical reasoning, there [is] no such time as "before the big bang."

If you claim the universe came from nothing, as my esteemed colleague Lawrence Krauss does, doesn't that mean there was some sort of nothingness sitting there waiting to make a universe? No, it doesn't. The true meaning of the universe springing from nothing is that there was no place, no time, no *thing* prior to the origin of the universe.

I have to say that in the last twenty years, opinion among cosmologists has swung to the idea that maybe reality is after all eternal, and that what we've been calling the universe all along is nothing of the sort. It's just a fragment, a component. It's one bang among many scattered throughout space and time, together forming a multiverse, which is eternal. So you get the best of both—you have an eternal multiverse, and you have individual universes with births, lives, and maybe deaths at the end.

That's the currently fashionable view, but to be honest, we don't know. Fred Hoyle, incidentally, liked the idea. He accepted the big bang, but he was a skeptic of it. He thought the universe was eternal, had no beginning, it will have no end, and that as it expands, new matter would be created to fill in the gaps. Evidence shows that that is really not the case, but it was an early heroic attempt to solve the problem of the origin of time.

Let me now move on to this subject of time travel that everybody's fascinated by. Could we travel in time, in principle is it possible? And my answer: not only is it possible, we've done it. We've done it already. In fact, we're doing it all the time. Traveling in time is actually very simple to explain. But, and maybe this is Bergson's problem, we tend to have this idea that there's the time, a type of a common time throughout the universe. What Einstein showed is that time is relative, that your time and my time are not the same, because we can get out of step. How? We can get out of step in two ways.

One is by motion. Jimena already mentioned this, often called the twin effect. If I get on a plane and I fly to London and then come back again, my time and your time have [gotten] out of step by a few nanoseconds, which is easily measurable with modern clocks. This is true. In fact, the global positioning system relies fundamentally on this time warping, or time shifting, effect that Einstein predicted way back in the early part of the twentieth century. So, high-speed travel is one way that different observers' times get out of step.

The other way is gravity. Time runs a little bit faster on the roof than it does in the basement. And again, you can test that with clocks. To get a bigger effect than in a building, you put a clock in space, compare the clicks with the clock down here, easily measurable, and they get completely out of step. The GPS wouldn't work if it didn't factor in both of these time-warping effects. So we know that they are real.

In both cases, the effect puts you (slightly) forward in time. Imagine a pair of twins, one who sits at home and the other who travels. The traveling twin, or the twin who goes into a deeper gravitation well, is also traveling a little bit into the future of the stay-at-home twin. On our current state of technology, we're talking nanoseconds here, not enough to make a Doctor Who–type adventure. But by traveling much faster, close to the speed of light, you could leap years and years into the future. You could reach the future quicker by traveling close to the speed of light or by going close to the surface of a black hole. At the surface of a black hole, time literally stands still relative to a distant observer.

We know these time-warping effects are real. We can measure them. For example, if I go off at very close to the speed of light to a nearby star, come back again after two years my time in the rocket ship, I may find twenty years [have] elapsed here on Earth (That's a specific example of the so-called twin effect. I'm just using arbitrary numbers in my illustration. You can have any number you want). In effect, I will have leapt eighteen years into the earth's future, but, and this is the big but, you can't go back again. You can't repeat the out-and-back journey in space and thereby get back to the time you left.

To go backward in time is much more problematic, but there's nothing in Einstein's general theory of relativity to forbid it. There may be other aspects of physics that rule it out, but our best understanding of space and time is the theory of relativity. That theory says, in principle, traveling back in time is possible.

Many of you will know, when Einstein fled from Nazi Europe, he made a home in Princeton at the Institute for Advanced Study. He used to say the only reason he went to work was to walk home with Gödel, the Austrian logician. Slightly crazy character, very eccentric, he was a logician and a mathematician, but he was captivated by the idea of being able to travel back in time.

He found a solution for Einstein's general theory of relativity involving a rotating universe that permitted an observer to go off, not just to anywhere in the universe, but to any when. He or she could go back and observe their own past. Einstein was deeply troubled by that. He really didn't want unrestricted time travel of that nature in his theory. But it's a perfectly good solution, and it has not been ruled out.

There are other solutions too, involving wormholes and whatnot. I'm not going get into all that tonight. But certainly, Gödel himself was very excited by the prospect of time travel. Freeman Dyson told me that Gödel would often come up to him and say, "Have they found it yet? Have they found it yet?" He was referring to the rotation of the universe. Because if the universe is rotating, you could do this time thing.

Now, I'm running out of time, so I'll go at a fast pace. I wanted to tackle the story of the arrow of time, because that's where we get into the most vexed territory. We have this everyday, overwhelming impression that time is flowing or moving or passing, there are many metaphors, a river of time, a flux of time, a whirligig of time (as Shakespeare described it). The passage of time is something that is absolutely fundamental to the way we perceive time.

To elaborate, we think of there being a present moment, a now, and that this now is somehow marching forward into the future. This impression is overwhelming. The problem is that in physics, it makes absolutely no sense whatsoever. I'm going to stick my neck out and say I think, along with Einstein—Jimena already quoted him—the past, present, and future are illusions. If that's correct, time does not pass. In fact, I claim it is meaningless even to talk about time passing. As Marcelo said, it does seem like time is speeding up, because it used to be an awful long time until Christmas when I was a child, whereas now I say to my wife, "It's Christmas next week, I haven't made any plans!"

And so we talk in this sort of flowing time language all the time. But it makes really no sense. Because how fast does time pass? I'll tell you: one second per second. The idea of something flowing or moving only makes sense if you have another time against which to gauge it. We can talk about how fast a ball moves: one meter per a certain interval of time. But time itself can't move or flow.

I think I know where this confusion comes from. When we say, "All my life it seems to me that time is passing, and this present moment is a special one," we're assuming that the self, our personal identity, is conserved. To be sure, I may change my opinions or my tastes, but it's still the same me that it has been from birth. That leads to the erroneous impression that a fixed me experiences different moments of time as time is flowing.

It's back-to-front, though. What's changing is not time, it's the me that's changing. Tomorrow's me is slightly different, only very slightly, from yesterday's me. My conclusion: time itself is just there. The me is what changes. Later me is different from earlier me. Each me at each instant is correlated with the state of the world at that instant. We mistakenly think that this is time flowing or moving. Time doesn't. Time, like space, is just there. You're at a point in time and space like you're on a point of the map of the landscape. Physicists just think of space-time as simply there, nothing happening. There are later moments and earlier moments, but no movement or flow.

There's certainly an arrow of time, that is, a distinction between earlier and later. But that's quite a different thing from the flow of time. A lot of literature, and a lot of philosophy—I have to call it bad philosophy—conflates these two meanings of the word arrow. The arrow is a double metaphor. It can mean you fire an arrow and off it flies, it's a metaphor for motion. But there's also an arrow on a compass needle or a weathervane. In that case it doesn't move, it simply points in a certain direction.

When we talk about the arrow of time, the correct metaphor is the latter. The arrow points toward the future, it's not moving toward the future. So people should stop using this term arrow of time, which has a legitimate physical meaning, to describe the illusory flow of time.

How, then, does the sensation of time passing arise? I liken the illusion of the flow of time to dizziness. I could give a demonstration. If you stand up and you twirl around like this, you get an overwhelming sense that the universe is rotating. It's not rotating. I can see it's not rotating. But it feels like it's spinning round. In the same way, it feels like time is surging forward. But when I stop and think about it, it can't be. It's myself that is changing.

I think the illusion of the flow of time has something to do with the way the brain works, our memory and so on.

M GLEISER: Are there any questions you'd like to ask each other before I ask you questions?

J CANALES: Yes, just one question about when you say that we are changing, and that it's not time. What is changing you? Are you saying there's a hidden something else behind it?

P DAVIES: Right. Because we have this sense of self. We are under the illusion that these selves are a conserved entity. And although the selves may experience different things, including about one's own personality, there's some inner core of me-ness which is unchanging.

But really, selves are just the sum total of all our experiences and introspection and so on. And every day, it's a little bit different. So the technical way of expressing this is [that] there is very high mutual information between today's Paul Davies and yesterday's Paul Davies. Less mutual information between the five-year-old Paul Davies [and today's Paul Davies]. These are different entities, and they have different mental states, and these mental states are correlated with physical states of their world at that time. At each instant of time there is a mental state and a physical state, with a Paul Davies in 1946 and a Paul Davies in 2018. And hopefully some (slightly different) Paul Davieses in the future, too!

But that's all there is. Everything about the world could be expressed in time-neutral terms. We could make no reference to time passing or flowing. Every statement that you make that refers to the flowing or moving or the passage of time could be recast in rather boring, unpoetic language without making any reference to that. Yet even some physicists don't believe that, and many philosophers don't. I was strongly influenced by J.J.C. Smart, the Australian philosopher who did more than anybody else to demolish the idea that time is moving. But the tradition in physics really begins with Einstein.

I'd be interested to know your point of view. Of course, in daily life we all feel time is moving. But as a philosopher, when you examine that, what do you find?

J CANALES: That was at the crux of the debate between Einstein and Bergson—Einstein promoting this idea of the block universe and Minkowski and Hermann Weyl adopting that, and that being, as you say, very absurd when we think about our daily lives. And Bergson had the opposite view. Time was heterogeneous. It was the eruption of the absolute new. It was indeterminate. It wasn't just a recycling of bits. You could not have a statistical thermodynamic interpretation of the arrow of time and history because that's just recycling. For him, time was about the bursting forth of novelty into the scene of life.

Bergson was a very arrogant philosopher. He looked back and he claimed that before him, for the previous one thousand years, everybody had been talking about things being static and fixed. Yet our daily lives, our arts such as poetry and music, our concerns, our feelings—so much in our world is completely motivated by the opposite. It was a very simple statement. He had to go back to the Platonists—to Plotinus, to Heraclitus—to find some allies in Greek philosophers.

He wrote a book against Darwinian evolution because the way that it was described by Herbert Spencer was just the recycling of the bits of organic matter already there, it wasn't real, creative evolution. And Bergson missed this aspect of science. For him, it was wonderful. It was useful. It led us to create these wonderful machines. It led us to be able to predict astronomical events, such as eclipses, but it did not explain our basic reality, or the things that really mattered to us.

In my book, I don't take sides for the block universe or for a Bergsonian view. In fact, when quantum mechanics came back with indeterminism, Bergson and some of his allies said, you know, it wasn't Bergson [who] was wrong on the science, it was the science that was wrong on Bergson at the time.

This is even a debate among scientists. There are contemporary scientists, some of your colleagues—Lee Smolin, Adam Frank—[who] have found Bergson and some of the continental philosophers who follow him inspiring.

Just to conclude, my point of view is that this is a real tragedy, and that it explains the twentieth century. It explains this division between the sciences and the humanities, between the arts and the sciences, between the world of experience and the world of rationality. It is representative of a divided century. A century that has been very violent, and a century in which there are large groups of people, large intellectual spheres, that simply don't talk to each other. There are silences in universities between the humanities and the sciences and the arts.

Even our own experience is in a sense divided when we talk about these things, when we think about them. We jump into certain registers, "Now I'm talking about objective time and now I'm talking about how we feel time." In movies, the cinematographic machine embodies this contradiction, in the sense that frames are all equidistant, they have the same timeframe, but the way that the stories are told jumps back and forth; there are flashbacks, there are premonitions. There's hardly a novel that is strictly chronological. In fact, there's hardly a science book about time that is chronological. That's more than ironic.

Regarding this division, instead of siding with one sense over the other, my point of view is to see how it has been representative of the twentieth century as we have come to know it, and to hope that perhaps we can move to a different type of dialogue.

M GLEISER: There's a complication that came up after the [1922] debate, which is the discovery that, as Paul was mentioning, the universe itself has a history. There is a sense of a cosmic time, which is the cosmological time, which is really a universal clock, in a sense. You can measure the distance between two galaxies, and they will move apart if they are sufficiently far apart to begin with. An arrow of time is implicit in this motion.

P DAVIES: But you need to explain that at any given point in space, there is just one reference frame that corresponds to that cosmic time. That's the frame in which the observer views the cosmic background heat radiation to be uniformly distributed. Now the times at each point in space are relative to a unique reference frame, but it's different from the reference frame at another cosmic location over there. But wherever the observers are located, [as] long as they are in the reference frame for which the cosmic microwave background is isotropic in that frame at that location, they will agree with other cosmic observers on statements like "the big bang occurred 13.8 billion years ago."

M GLEISER: Definitely. But that's kind of interesting, because then that tells you that the universe itself has changed, and you can tell its story.

P DAVIES: Oh yes, absolutely. There is an arrow of time, which is certainly objectively there. Nobody would say that the universe is unchanging. Later states of the universe are demonstrably different from earlier states—it's bigger now, it's cooler, the entropy is greater. There are all sorts of ways in which the universe is changing. But nothing is flowing or moving or anything of that sort.

The most distinguished scientist that I ever met who was convinced that the passage of time is an intrinsic property of time itself, not a property of the world, was Ilya Prigogine, but he was a chemist. We agreed to differ.

J CANALES: And he and Isabelle Stengers wrote about the Einstein-Bergson debate.

P DAVIES: Oh, that's right, they did, they did.

M GLEISER: Do you want to mention something about time and matter asymmetry? Because I think that's very important, and we haven't talked about it.

P DAVIES: There's a theorem, which had better be true in almost any sensible fundamental theory of physics, called the CPT theorem. What that means is when we talk about the arrow of time, and we're thinking of an arrow of direction (not flow!), if you flip it, that's time reversal, an operation that is often denoted by the letter T. Applying T to the laws of physics, wherever time T appears, you get—T. The other one, C, stands for charge. And that means if you've got a positive charge, the C operation flips it into a negative charge. So a proton would turn into an anti-proton, and an electron into a positron. Finally, P stands for parity, that's right-handed and left-handed. So, in a mirror, do the laws of physics look the same or not?

The CPT theorem says if you combine all three operations together, then collectively, they must be respected by nature. If you reverse charge, parity, and time, everything should be the same. That is the law of the theorem under the combined CPT operation, but not necessarily under individual components of that. The matter/anti-matter asymmetry refers to the CP bit. Therefore, given that CPT is conserved, if T were to be violated, then CP would be violated. And that is in fact the case. Experiments to test both T and CP show that those symmetries are very slightly violated in nature.

That's a very important discovery. It's one of the big mysteries about the universe: where is all the anti-matter? We can make matter in the lab easily. It's done every day. But whenever you make matter, you make an equal quantity of anti-matter. You make a proton, you make an anti-proton; you make an electron, you make a positron, and so on. So, if the big bang turned energy into matter, and we're made of this matter, what happened to all the anti-matter?

The feeling is that there must be an asymmetry between the laws of matter and the laws of anti-matter, that in the early days after the big bang there was almost an equal quantity of both, with a slight preponderance of matter. Then the matter and the anti-matter annihilated and created all the photons that gave us the heat of the universe, and this leftover residue of matter is what the stars and planets are made of.

But that goes with a concomitant reversal of time. I said at the outset that the laws of physics are symmetric under time reversal. That's true of the laws that govern electromagnetic waves. And it's true of almost all everyday phenomena. But it is not true of some sub-atomic radioactive phenomena, namely the weak interaction. There is a slight breaking of the symmetry between time forward and time backward, which shows up in very careful particle physics experiments. It doesn't affect the fact that when you stir your cream in your coffee, it mixes up and [if] you keep stirring, it doesn't unmix. All that's left alone.

So at the level of fundamental particle physics is this very curious type of asymmetry, without which we would not be here. Because it must be that, in some way that's still not completely worked out, this CP matter/anti-matter symmetry breaking yields a universe of matter only.

M GLEISER: Jimena, you mentioned authority and who has authority, and that was one of the most important themes in your book. So, who has the authority to define time, and why?

J CANALES: The debate between Einstein and Bergson was, historically speaking, the moment in which we see the reversal, and physicists start taking on that role. And that role has only accentuated. Nobody would think to go to an artist or a poet to find the truth about time. The role of the philosophers and the humanists has been sidelined more than ever.

An important paper was Hilary Putnam's, a philosopher who was my teacher at Harvard. He wrote a paper in 1967 where he basically repeated the point that philosophy had nothing to say about the problem of time. A lot of things that he wrote afterward contradicted that early statement, but it was a very important claim at that moment.

And we have continued comments in the public sphere, from Neil DeGrasse Tyson to Richard Feynman, that continue to claim that philosophy has nothing to say about this issue. Feynman said that philosophy was as useful to physicists as ornithology is to birds. I think it's pretty clear that if you have a seminar about the nature of time, you're going to have a physicist like Paul Davies. That's who, as a society, we have given the authority to talk about the nature of time.

What's interesting for me about that is that the role continues even when there's a clear contradiction in the time-telling techniques used by ourselves in our narratives, used by filmmakers, used by novelists. What can we do with that division? That's what I find interesting.

P DAVIES: In defense of philosophy, just last week I went to a talk by Frank Wilczek, one of our colleagues here in the physics department, called "Plotting Mysteries of Time." He briefed the listeners that the last word has not been said on all fourteen aspects of the nature of time. I think that physicists can give a pretty reliable account of all sorts of things about the nature of time, and there are experimental tests that we can do, as I've already mentioned. But there are still open questions.

In spite of my evangelizing about [how] time does not pass, I still don't feel entirely comfortable. If the flow of time is an illusion, it's an illusion that we can't give up. Like free will; very many scientists and philosophers don't believe in free will, but we can't afford to give it up, otherwise normal life becomes impossible.

M GLEISER: Time for some questions.

QUESTION ONE: A physics professor explained the direction of time as a consequence of the law of entropy, that you can't create order from disorder. And that always satisfied me intellectually. Is there a reason I shouldn't have been so easily satisfied by that explanation?

P DAVIES: There are two reasons, I think. But I'll try and keep this brief. First of all, steer clear of talking about the second law of thermodynamics.

C. P. Snow cited it as an example of the chasm between the humanities and the sciences. Sitting at High Table in Cambridge you would be ashamed to say, "I've never heard of William Shakespeare," but if you said "I've never heard of the second law of thermodynamics," that's OK.

The second law basically says that the universe is getting more and more disordered or chaotic over time. And that's what gives us one reason why there's an arrow. There are other arrows as well, but that's the most pervasive one. And so, the simple statement is that the universe started out in an orderly state and becomes more disorderly over time.

That leads back to the question of why did it begin in an orderly state? That puzzle is further compounded by the fact that when you look at the early universe, it doesn't seem to be an orderly state. It seems to be a maximum entropy state, a state of thermodynamic equilibrium. What was out of equilibrium was gravitation. And the gravitational aspects of entropy have still not been worked out. It's still very much an open project.

Then there's another issue, why it's not quite so simple to say entropy did it, which is, as John Wheeler used to like to say, "if you ask an atom the direction of time, it will laugh in your face." He had this wonderful turn of phrase. What that means is at the level of individual atoms, there is no time forward or time backward, it's just the behavior of large numbers of things.

Just to give you a simple example, if you take a deck of cards, which is newly purchased, it would be in numerical order. If you shuffle it, it will become disordered as a result. And then you think, if you keep on shuffling will you get back to that original state of order? In principle, yes. You'd have to shuffle for a very long period of time. But if you look at an individual card, the notion of order or disorder is irrelevant. It's relevant only in relation to the collection. In that sense, it makes the arrow of time a somewhat subjective concept. This has to do with our view of the world—the fact that we don't see individual atoms. It's Laplace's demon. ["We may regard the present state of the universe as the effect of its past and the cause of its future. An intellect which at a certain moment would know all forces that set nature in motion, and all positions of all items of which nature is composed, if this intellect were also vast enough to submit these data to analysis, it would embrace in a single formula the movements of the greatest bodies of the universe and those of the tiniest atom; for such an intellect nothing would be uncertain and the future just like the past would be present before its eyes.— Pierre Simon Laplace, *A Philosophical Essay on Probabilities*]

It's an interesting theological issue—if God is all-knowing, is omniscient, knows every atom, then God has no arrow and no flow of time. We know something God doesn't. That's one for the theologians.

QUESTION TWO: I have a question for the historian. I was intrigued by your comment on Einstein's view of philosophy at that Bergson debate, because his 1905 paper on the electrodynamics of moving bodies seems to me to be entirely a philosophy paper. He makes no reference to experimental data and is motivated entirely by a very aesthetic imbalance in the laws of electricity and magnetism. Did he change his views over time? Was he aware of this change in his own behavior toward the more philosophical questions?

J CANALES: That's a great question. Einstein definitely had an implicit philosophy, and most scientists have an implicit philosophy. And the implicit philosophy of most scientists is materialism and reductionism. When you read some of the scientists [who] want to push philosophy away, they themselves have a philosophical view, but it's implicit. If you study philosophy, if you study history, you can place them within certain philosophical schools.

Einstein's reaction against philosophy at the time was really a reaction against the authority of Bergson and the authority of other people who previously had the right to speak about time and who were the go-to persons for thinking about it. During the debate, he said things that he would never say again. For example, he would say that his definition of time was the only objective definition. Then later on, he would continue to say that metaphysics had no place in science, that philosophy was subjectivist, and that there had to be just technical answers. But as he grew older, he actually recanted and changed that position.

At the beginning, he was also very inspired by the philosophy of Ernst Mach, which was very operationalist, very much based on actual sensations. Then he said that it was sterile. He thought that the logical positivists, some of the philosophers who had actually jumped to his defense and used relativity as an example of the best type of science that can be done, had pushed away metaphysics, and he said, no, we actually need metaphysics. He thought that the difference between physics and metaphysics was actually of degree and not of kind, at the end of his life. So he definitely changed.

P DAVIES: The later Einstein and the earlier were not the same person.

QUESTION THREE: The Greeks had this concept of *kairos* and *chronos*. Is the distinction between Bergson's duration and Einstein's time sort of the subject and object difference?

J CANALES: That's how it was broadly understood. Bergson was seen as defending a subjective notion of time. But Bergson denied that characterization of his own philosophy. He was interested in the *point de capiton*, the anchoring point [that] ties the objective and the subjective together. He thought that it was absurd to think that there was just something objective.

For example, when Einstein defined time by a clock measurement—which makes a lot of sense. We have clocks, time is what clocks measure. Seven o'clock is the pointing of the needle of my clock to seven and the train arriving at the station—Bergson responded, "Why are clocks made? Why are they bought? Because we have a prior notion of time. We care about events. We want to go places." He thought that it was therefore absurd to define time via clock time.

The other interesting thing is that when you look at Einstein's correspondence and his private writing, he's obsessed with time passing by in the same way that Bergson describes it. He's a person who's always running out of time. He's really, really busy. It's really an obsession. In certain years, one sees this obsession in pretty much every letter. He calls it the old tune, the old refrain, that he doesn't have time. So inside Einstein himself there is that contradiction between the objective and the subjective.

QUESTION FOUR: I've always construed time as something that very much needs to be defined in terms of other quantities, like velocities of particles, or even just motion itself, broadly. But it seems like we don't need to do that for fundamental quantities like charge. In your opinion, is time fundamental, or is it better understood as a useful fiction?

P DAVIES: Very good question. We certainly treat it as fundamental in, for example, Einstein's general theory of relativity. Space and time are just there. Space-time is not an arena, it's part of the drama of nature. It has its own dynamics, its own gravitational waves that propagate in empty space-time. It's simply like the substrate of the universe, so you take it as a fundamental quantity.

But there's a long tradition of people who feel that time, indeed space-time, might be an emergent property. Proponents of this include John Wheeler, who had vague ideas of pre-geometry. The analogy he used was a block of rubber. It looks continuous and it can be warped, just like space-time. But if you go down to the atomic level, you'll see that there are atoms that are bound together, and the notion of a continuous medium breaks down. In the same way, space-time might have some component, some elementary entity out of which space and time emerge.

Roger Penrose had something similar. There have actually been many, many attempts to have a "space-time coming out of . . ." theory. String theorists wanted space-time to emerge from the physics of their theory. That hasn't happened yet.

What do I think? I think that probably yes, we need a more fundamental concept out of which we build space-time, matter, and indeed, I would even say, the observer. We need to somehow incorporate that. Which surprises some people.

QUESTION FIVE: Professor Davies mentioned that there was nothing before the big bang. But there was something that caused it. And the thing that caused it must have taken some period of time before causing it. So what do you mean by that?

My second question is, if time doesn't move, then what does every single tick on my clock suggest?

And the third question is, as you said that time is static, it doesn't move, then can't we imagine time to be a spiral track on which we or the world moves? Because I come across 1 a.m. every single day. But that 1 a.m. is not the same as the 1 a.m. that I had yesterday.

P DAVIES: Right. The simplest one is that clocks measure intervals of time, they don't measure the flow of time. The clocks are really about how long it is between this event and that event, but it's got nothing to do with flow. So that's what a clock is.

As far as causation and [where] the word *cause* is concerned, in normal, daily life, effects follow causes. If we're talking about the cause of the origin of the universe, it's rather meaningless if there is no time. So the better question is, how is it that the universe can come into being? How is it that space-time can come into being from nothing?

If the laws of physics permit it, then that's an explanation. It's not a cause, it's an explanation. But then we're driven back to ask, are the laws of physics within time, or do they transcend space-time? Did they come in with the universe, or do they have, like mathematics, an independent existence? I've written very extensively on this, carefully unpicking all of this stuff. Made it a book called *The Goldilocks Enigma: Why Is the Universe Just Right for Life?* Please may I refer you to that book where all of these aspects are discussed very carefully in easy language? Because people get in a terrible muddle about words like cause and effect.

QUESTION SIX: How do you think the perception of time relates to the size and metabolic rate of the organism, or the observer of the time.

P DAVIES: There is a scaling law. Little things live fast and die young, like the mouse, animals like that. Elephants are more ponderous, and humans are somewhere in the middle. You get a certain number of heartbeats. The fact that the mouse is scurrying around—is it seeing everything like a fast-forward movie? Of course, we can't know.

But what we can know, and here we're getting more into physiology and the neurosciences, is response. I once met a man called Miroslav Holub who wrote a book called *The Duration of the Present Moment*. He was interested in music and the way we sort of integrate it all in the brain. We don't dissect it. Somehow your consciousness is spread out over, I think he said it was 1.3 seconds. So forget your tenth of a second, it might be something longer.

But if you look at a tennis player, you know the speed of the ball, there's no way they could think, "Well the ball's going fifty miles per hour so I'll step out now and do this." You can't. It's all happening much faster than the speed of thought. Somehow this perception we have of, if you like, the flow of time, or at least the succession of events, is a narrative that the brain reconstructs after the event.

We come back to this idea of thinking of ourselves as unitary entities—a self that's merely inside, with all this sense data coming in. This conversation is what I'm conscious of at this moment. It's a conscious experience at this moment, but it's all a fiction. When you do the experiments, what you find is that the brain is editing various drafts, it goes back and sometimes reverses the order of events.

You can do experiments where you ping somebody's brain to make them feel that one finger's being pricked, then you prick the actual other finger, and you can show that the conscious subject reverses the order of the experiences. There's an awful lot of editing going on inside the brain. So heaven knows with a gnat what other things are going on. Can it make split-second decisions? Could it play tennis better than a tennis player? I don't know, there must be some people who do.

QUESTION SEVEN: Regarding whether or not there's a flow of time, is there some experiment that you could do?

P DAVIES: What could there be? Supposing you woke up tomorrow and somebody said, "A terrible thing happened in the night—time doubled its speed." It's like saying, "We're all twice as big, but so is the universe."

You have to relate it to something else. You have to say, "Here is something that measures the speed of the time we have now relative to some other time that we haven't yet discovered." We could do that. There was a book written by somebody called J.W. Dunne many, many years ago called *An Experiment with Time*, in which he said there must be another time to time our time against. But then maybe there's another time for that and another time for that. So you get into an impasse there. I don't think these ideas are very helpful. I come back to my claim that what's changing is not time, it's us, basically.

6

CYBORGS, FUTURISTS, & TRANSHUMANISM

A Dialogue Between a Neuroscientist and an Author

MGLEISER: This big question is about transhumanism and the blending of flesh and machine. How far down this road are we? How far can we go? And, perhaps most importantly, how far should we go with this? Clearly, these are questions that call for a scientific and a humanistic outlook; they may represent interesting, even amazing scientific challenges, but they also directly impact our human identity and, why not say it now, possibly the future of our species as a whole. Is the marriage of carbon and silicon our evolutionary fate? The new technology also has practical impact on the job market, as AI and robotization will push many out of their current jobs while creating the need for new kinds of experts and of technical training. The topic has the potential for endless speculation.

Ed Boyden, one of the leading neuroscientists working at the cutting edge of this topic, is the head of the Synthetic Neurobiology Group and an associate professor of biological engineering and brain and cognitive sciences at the MIT Media Lab and at the McGovern Institute for Brain Research. Ed is really trying to make headway in some of the most difficult questions concerning the notion of how the brain engenders consciousness, what that means, and whether we can reproduce this process (if that's what it is) artificially or not.

Mark O'Connell is a journalist and humanist who has addressed these questions in a book that explores and meditates on transhumanism and what it means to us as individuals and as a species, *To Be a Machine: Adventures Among Cyborgs, Utopians, Hackers and the Futurists Solving the Modest Problem of Death*. In 2018, the book won the UK's very prestigious Wellcome Book Prize.

E BOYDEN: I direct a group at MIT that works on brain technology—ways to map what's happening in the brain and to try to repair what goes wrong. Over the last century, in neurology and psychiatry studies of patients who were injured in war or who had lesions from disease, we've learned a lot about how different regions of the human brain are important for decision making, for memory recall, and for emotion. For example, patients who have an injury to one part of their brain lack fear, whereas those with changes in another part of the brain are no longer able to form memories.

In the twenty-first century, we're now finding that there are many new technologies that are having huge impact on our ability to understand and repair brain functions. So I thought today I would give you a bit of a lightning tour of just some of the areas where we are seeing very rapid change, and then talk about the consequences on topics like cyborgs and transhumanism. Can we simulate what happens in the brain? Can we enlighten ourselves by understanding more about the human condition? Can we augment the brain?

The human brain is incredibly complicated. There are around a hundred billion brain cells called neurons and they're connected, each of them, to maybe a thousand others. They compute using electrical pulses and they communicate using chemicals. If you could scan a human brain, one human brain, with enough resolution to see the finest building blocks of the brain, individual biomolecules, and you stored that data on little hard drives and stacked them up, that tower of hard drives—remember, for one human brain—would reach way into outer space. We're talking about incredible complexity.

Furthermore, we often don't know what we need to look for in the human brain, or in any brain for that matter, in order to capture the essential information processing that allows us to compute thoughts, feelings, sensations, and actions. For example, just in the last decade or two, people have found that brain cells make cannabinoids, molecules that have effects not unlike the active ingredient in marijuana. These molecules may perform computational functions that we don't fully understand. If you wanted to make a map of the brain, how much information do you have to acquire? The answer at this point is [that] we don't know for sure.

Another question that often comes up when we talk about cyborgs or transhumanism is if you could actually simulate a brain in a computer, would that be you? Of course [some] of the biggest questions of all [are] what is experience, what is consciousness, what is subjective feeling? At the current point in human history, we cannot create consciousness directly or sense it directly, and so this remains a mystery.

Let me tell you about a couple of technologies so you can see the frontiers, and then we'll get into some of the more futuristic directions.

If you look at a brain scan, a magnetic resonance image, of a human who's thinking, the blobs that you see lighting up are measuring blood flow. Blood flow changes in the brain indirectly as a result of brain activity. The problem, sort of one of the dirty secrets of our field, is that if you look at these blobs, each of these blobs contains millions, maybe billions of different brain cells, and they could all be doing very different things.

I tried with that [example] of the tower of hard drives that reaches into outer space to give one perspective on the complexity that we're dealing with. But the size scales that we deal with are just tremendously, vastly daunting. If you could somehow take a single human brain and swell it to the size of a city block, the connections between brain cells, known as synapses, would be the size of individual grains of sand. And those connections are full of molecules, molecules that generate electrical pulses, molecules that exchange chemical information. And every one of these connections, in principle, could be different, which is one idea of how information is stored in the brain.

So one question that we and others are trying to confront is can you actually map the brain across these vastly differing spatial scales? Brain cells are enormous. They can be centimeters in spatial extent. And yet these minuscule connections are so incredibly tiny.

One idea that we've been working on is borrowed from physics. This goes back about forty to fifty years to the study of so-called responsive polymers. That's a fancy way of saying the material in baby diapers. What happens when a diaper is used by a baby? I guess some of us have experience with this. Water causes these polymers to swell. These polymers absorb the water through osmotic force and they swell up and become bigger.

One thing that we've been working on is to use this technology to physically expand brains so that you can map the finest connections between cells and the molecules at those connections. What if you could chemically weave these threads of polymer around biomolecules, in between biomolecules? And if you did it just right, when the diaper material swells, you'd pull all the building blocks of life away from each other until they're far enough apart that you could take a picture of the brain, in all its glorious detail. Maybe even using really cheap cameras, like the kind you find on cellphones.

We spent a lot of time figuring out the chemistry, using mouse brain tissue (Mice are very commonly used in neuroscience as a model, since we can't do everything we want on human brains. This of course only works on preserved brains. Expanding a living thing doesn't really work). We formed the baby diaper–like polymer throughout the piece of mouse brain tissue and then added water. In a time-lapse movie sped up about fiftyfold, you can see this piece of brain tissue growing before your very eyes. We can swell specimens a hundredfold, ten thousand times even, in volume. So now we bring the very fine connections between neurons, and the molecules at the connections, into view because we can separate them until you can actually tell them apart.

Could you take these cells, these tissues, these circuits, and have a constellation of molecules hovering like stars in the sky, except they [are] anchored to sites on the polymer mesh that pulled them apart? The idea here is maybe we can make maps of the brain so fine that you could try to simulate how they work in a computer. Using this kind of technology and combining it with clever chemical techniques that many groups are developing, like ways of color-coding neurons so that they all look different, could you make a map of the brain accurate enough that you could simulate a brain computation?

That's technology number one, and before we get into all the implications of this, let me tell you about technology number two. Everything I've shown you with expansion only works on preserved tissues, which is a euphemism for dead tissues. But the high-speed electrical pulses that brain cells generate and the high-speed chemical exchanges between neurons, we need to observe them in the living state, and ideally, be able to control them. If you could activate brain cells, you could try to figure out what can they initiate in terms of behavior or decisions or emotions. Or by turning off brain cells, you could figure out what are they needed for. Could you delete a memory temporarily or eliminate an epileptic seizure or a Parkinsonian tremor?

To do this, we've been working on ways to equip brain cells with, effectively, solar panels—molecules that convert light into electricity. The brain doesn't feel pain, so you can bring optical fibers into the brain. Over a third of a million human subjects have already had electrodes implanted in their nervous systems for conditions like Parkinson's disease and deafness. But instead of an electrode being inserted into the brain, we can deliver pulses of light down an optical fiber.

Now the tricky part, of course, is that neurons mostly do not respond to light, with the exception of those in our eyes. So to make this technology, which we call optogenetics, work, we had to go to single-celled microbes like bacteria and algae. Single-celled algae turn their flagellae, those little tails coming out the back, based on signals from a little organelle called an eyespot. If you zoom into the eyespot, you can find molecules that act sort of like those solar panels. You shine light on them, of the proper color—in this case blue light—and they open up little pores, or holes, that let charged particles—ions—pass through. And that's exactly what we want. The resulting electrical signals are similar to those in neurons, brain cells, when they compute ordinarily.

If we take this molecule and put it into a brain cell and then shine light on the cell, light hits these molecules, opens the pore, let's the ions in, and activates the brain cells. In the same way that neurons in your brain are responding while I say these words.

We were very lucky. It turns out that this is a protein encoded by a small gene and we could take the gene and, using the technologies of gene therapy, a very exciting field, we could transplant the gene into a brain cell. Then the brain cells amazingly—this is sheer dumb luck frankly—serendipitously manufacture the proteins and, even more remarkably, install them in the right place. Lo and behold, we found that we could actually use these molecules to mediate the transformation of light into a meaningful neural electric signal. People are using this technology for all sorts of studies to understand the brain, and the resulting discoveries have even led to a couple of companies exploring new ways to treat brain disease.

Let me give you an example from my own life, which I think exemplifies how you can use these tools. I have a close collaborator, Li-Huei Tsai, at MIT, and she's an expert on Alzheimer's disease. Her group used our technologies in a study to try to figure out what activity patterns in the brain you could drive that might heal an Alzheimer's-afflicted brain.

Her group took our molecules and put them into a relatively rare population of cells in the brain and drove them with light at a special frequency—forty times a second. Forty times a second is kind of a magical number in neuroscience. For example, when you pay attention to certain things, the brain regions that are involved will increase their forty hertz, as it's known, activity. Lo and behold, what her team found was that in Alzheimer's model mice—these are mice engineered to have mutations similar to the ones that can cause Alzheimer's in humans—the immune system of the brain would turn on after the brain was driven at this frequency. It would start cleaning up the amyloid plaques and other molecular hallmarks of Alzheimer's disease.

What her team is working on now, based on this discovery, is how to induce these patterns of activity noninvasively. Of course, a gene therapy can be very precise. It's a great way of studying the brain. But gene therapies that have been making their way into humans are expensive and difficult to deploy. So with a collaborator, Emery Brown, our teams, led by Li-Huei, started experimenting with the idea of inducing these brain activity patterns through the eyes. Could you just watch a movie, and would the movie then induce these patterns of brain activity? And in the mouse models, that worked. So actually, Li-Huei and I have co-founded a company [that] is now doing human trials as we speak to try to figure out if you can build movies that would treat Alzheimer's disease. We're working on other technologies, as well, that are not invasive that would allow us, hopefully, to access deep structures in the brain.

Now, a lot of our motivation has been about brain disease. Over a billion people around the world, if you include stroke and addiction and lots of other conditions that relate to the brain, suffer from brain diseases. Brain diseases, I think, are particularly scary because they not only change our time to live, but they can change who we are, how we relate to loved ones, and so forth. Furthermore, pretty much no brain disease can be fully cured. As somebody who has a grandmother with Alzheimer's and just seeing, of course, the global burden of brain disease, this has been a core motivation for my work.

One of the goals of this evening is to delve a bit beyond the healing of pathology. What does it look like in a future where being able to simulate or augment or change how we exist as humans might be possible?

Let me talk about three areas I see in the coming five to twenty-five years as being urgent to think about. What do we want to do as a species in terms of the directions we want to head toward? What ethical considerations do we want to put forth that guide our thinking?

Let's start with simulation. Artificial intelligence is a hot topic nowadays. I think many of us have heard about these programs that can learn how to play chess or Go by themselves, and which have beaten some of the best game players on the planet. One of the exciting directions that I think is on the cusp relates to technologies like the one I showed you earlier, where you could expand brains and perhaps even map the finest molecular details of brain wiring. What if it [were] possible to make more brain-like artificial intelligences? As an example, artificial intelligences to date have been very good at playing games or solving problems that for the most part we kind of know how to do.

But what about making artificial intelligences that can be creative, that can solve real world problems, that exhibit attributes like ethics?

Would mapping parts of the brain be useful for understanding the algorithms that we use to make our decisions? As I mentioned earlier, a very common question is, would it be You if you were to map a brain and simulate it? As I mentioned earlier, we don't have any way of detecting or creating consciousness directly. In fact, you don't know for sure if I'm conscious, right? Although I hope you give me the benefit of the doubt.

This raises all sorts of interesting questions. There's a widespread fear of super intelligence nowadays. People have been publicizing that. There's also a lot of desire to boost our intelligence by giving ourselves greater powers to solve problems. The tension, I think, between this idea that we need to augment intelligence and this worry is an area of discussion that I think deserves more attention.

In our own group and with our collaborators, we have a network of many, many dozens of groups that we're working with. One of the goals that we're exploring is whether we could simulate in the coming, let's say five to ten years, a small brain, like that of a fish or a worm. To put things into context, a small worm like they study in many parts of biology might have a couple hundred brain cells, neurons. A small fish might have a thousand times as many, maybe a hundred thousand. A mouse brain would have about a thousand times as many neurons as the fish and a human brain has about a thousand times more neurons than the mouse.

So it's many leaps of magnitude that have to be achieved in order to simulate something as complicated as the human brain. The earlier [examples] that I gave, I hope, give a sense of how daunting the sheer amount of data processing will be.

On the other hand, you could imagine trying to build brains out of organic substrates. What if you could grow brain cells in a dish? For example, several companies have started to try to grow vegetarian meat. Could you grow meat in a dish without having to sacrifice the life of an animal? Maybe it's not so far off that you could grow realistic brain circuits in a dish. And if you could do that, what are the consequences—technological, humanistic, ethical and otherwise—that should be governing how we think about such things?

So simulation, I think, is an area where we could learn a lot about what the brain does. We could potentially generate new kinds of artificial intelligences, and I think the time is now to start thinking about the consequences and the path.

Second thought: enlightenment. My motivation for going into neuroscience was that I wanted to understand the human condition in scientific terms, which I know might not be possible, but it was at least one of my motivations for getting interested in the brain. One of the roles of our cultures and societies and religions is to help us understand how we can, as groups or individuals, overcome suffering.

It's interesting though, to see technology being used to look at questions that border on ethics and judgment and questions that are approaching what one might say are almost the domain of philosophy. I want to give you two examples from colleagues who are doing some very exciting work in this area.

One of my colleagues at MIT, Rebecca Saxe, had done an experiment some years back where they looked at people's judgments of other's ethics. As often happens in such studies, you give people stories, and you ask people to judge the characters in the stories.

Suppose Alice tries to poison Bob and Bob dies. Pretty much everybody thinks Alice is bad. If Alice tries to poison Bob and Alice is an incompetent poisoner and Bob lives, maybe she was trying to poison his coffee but accidentally used sugar, most people still think Alice is a bad person. What Rebecca and her colleagues found was if you stimulated a certain region of the brain, the right temporoparietal junction is a technical term, people would give Alice the benefit of the doubt in the second case. They would say, "Oh, Bob lived, not so bad." They value the outcomes over the intent. The way they did this was that they stimulated this brain region with magnetic pulses. You put a coil of wire next to the scalp outside the brain and you deliver a current to this coil of wire and it noninvasively perturbs the electrical activity in the part of the brain underneath.

This raises lots of questions. If the substrates that are making ethical judgments are that plastic, what does that mean in terms of what happens in our brains during our ethical judgments?

I'll give you a second example from Dayu Lin and David Anderson. They used the optogenetic technology that I described earlier to try to find sites in the brains of mice that would trigger aggression or violence. They found a tiny cluster of cells deep in the brain, [and] when they put our gene into those neurons and aimed light through an optical fiber into the brain, the mice would attack whatever was next to them, even if it was just a rubber glove. So here is suddenly an insight into a circuit that can create a behavior that, again, sort of touches on topics of ethics and judgment and even justice.

As technologies for observing and manipulating the brain become more and more common in science, I think an open question is: these technologies, which are also being explored for therapeutic uses; how can they potentially help us understand ourselves and improve ourselves?

Let me give you two quick examples. Giving a name to something can be very powerful. Look at the ongoing, obviously not complete de-stigmatization of mental illness. It's no longer thought of as a demonic possession or somebody's fault if they are mentally ill. It no longer becomes necessarily the defining part of one's identity. We can give ourselves distance from things by having names for things. One of my hopes is that we're eventually able to understand why we decide things the way we do. Would that give us more insight into our own behavior? Could we have more empathy, for example? Could we understand our motivations at a level that can only come through understanding mechanism?

Longer term, and this is something that I think is very interesting to discuss, is what happens with these studies where direct control over the brain is used to change the brain's function in a way that affects ethics or judgment or other kinds of philosophically relevant factors. We've all seen dystopian movies like *A Clockwork Orange*. Clearly there are wrong ways as well as potentially right ways to use such technologies, but the discussion, I think, needs to begin now.

My third topic that I want to address is augmentation. Augmentation is also a topic of great ongoing discussion. The use of attention-boosting drugs like Ritalin or Adderall in schools, for example. A lot of attempts to boost brain functions through training. In an era with technology that could potentially enter information into the brain directly, what does that lead to? As I mentioned earlier, over a third of a million people have already had neural implants, mostly deep brain stimulation for Parkinson's disease and cochlear implants for those who are deaf.

As technologies such as those are being used more and more broadly, the scientific community is finding that it is possible to do things like boost memory or change feelings. Deep brain stimulation, for example, has been explored for depressed patients. If you look at the trajectory that antidepressants have taken over the last several decades, they've broadened in usage until you might even consider some uses to be serving an augmentation purpose.

It seems that as humanity, we have a desire to augment at the right pace. If a technology is used to heal the sick and it's proven safe and effective, then, as the history of medicine shows, such technologies can broaden in usage over time, to the point where [they] can serve an augmentation purpose.

We can take a cue from other areas of biology that are more mature, such as the editing of the genome. In 1974, Paul Berg, a Nobel Laureate, called a meeting [in Asilomar, California], which was attended by the press. It involved scientists and lawyers and scholars and all sorts of different voices. And this is a period of great distrust in government. There were all sorts of fears about recombinant DNA, the idea of cloning a gene and transplanting it into a different organism. The people converged and said, you know what, we have to make some decisions here. If we're going to self-govern, if we're going to take these technologies forward in a world-impactful, positive, and ethical way, we have to talk about what we want to do, what we don't want to do, and what we might do. But we have to discuss the guidelines and principles to minimize harm and to make sure that stakeholders are represented.

It's been nearly half a century since, and one could argue that the impact of this conference has been enormous. In those decades, we've seen countless therapies emerge—treatments for hemophilia, human growth factors, all sorts of treatments, and sometimes even cures, that are made through recombinant DNA and gene cloning. Maybe it's time to have what you might call the Asilomar of the brain. The technologies that are already being used in science and that have already begun to be used in clinical trials—and sometimes, in the case of cochlear implants or deep brain stimulation, in approved forms in the population at large—are powerful technologies. I think we have to decide as a species, what do we want to do? Can we self-govern in our invention and deployment of such technologies, now that they're really starting to take off, in this quest to understand and help repair disease? And coupled to this trend of human augmentation that you see, can we get out ahead of the problem and make it into something that could be a powerful solution, potentially?

M O'CONNELL: That was really fascinating. Mind blowing actually. (Dying to get that one in.)

I'm going to start with two things. First, a disclaimer and then a definition. In the years since my book *To Be a Machine* was published, it's become clear to me that this is something that I should state up front whenever I talk publicly about the book, because sometimes I forget to do this and people wind up wondering occasionally when I'm going to get around to the pitch, to the bit where I talk about how we're all going to become immortal or merge with artificial intelligence. So here is the disclaimer bit for the avoidance of doubt: I am not now, nor have I ever been, a member of the transhumanist movement.

Now the definition, because I've learned not to assume that just because people are in a room listening to me talk on the topic [that] they necessarily know what transhumanism is. So transhumanism, to give you the brief but sort of functional gist, is a movement predicated on the conviction that we can and should use technology to push out the boundaries of the human condition. In the name of this Promethean ideal, transhumanists are committed to all sorts of strange and unsettling ideas and practices. Things like inserting microchips under the skin to give themselves superhuman extrasensory capabilities, to become de facto cyborgs, or like cryonics, whereby they freeze their bodies in liquid nitrogen shortly after death with a view to being resurrected at some point in the future when the technology becomes sophisticated enough to do so.

Transhumanists believe in a future in which they will achieve immortality by scanning their brains and uploading the data into machines. Kind of very extreme extrapolations of the kinds of things that Ed is talking about and working on.

The underlying principle of all of this, of the movement and all its various kind of enthusiasms and convictions about the future, is that it's our destiny as a species, that it's sort of necessary and desirable to transcend our flesh and blood bodies and to become something other and better than the animals that we are. I'll say more about this stuff as I go on, but that's sort of the basic definition, just so you know what I'm talking about, which is always useful.

As far out as it sounds and is, this movement and its ideas have firmly taken root in the soil of Silicon Valley, and there are powerful and influential people who are committed to bringing about some version of this future. The Google founders, for instance, are dedicated to solving what they consider to be the problem of human aging and, in 2014, set up a biotechnology research and development firm called Calico Laboratories, explicitly aimed at finding solutions at the level of genetics to the problem of aging.

Peter Thiel—who I write about quite a bit in my book—the venture capitalist billionaire [who] founded PayPal, has invested significant amounts of money in the project of achieving immortality through technology, most pressingly his own immortality, I think it's fair to say. Elon Musk, Thiel's former PayPal colleague, has spoken publicly about his conviction that the rise of artificial intelligence is something that will lead inexorably to our becoming obsolete as a species, that AI will evolve as far above us as we have evolved over the lower primates, and that the only way we can avoid this fate is by taking evolution into our own hands [and] developing the power of brain machine interfaces and merging our actual minds with AI.

So it's important to bear in mind that all of this stuff sounds crazy, and much of it is, in my admittedly fairly untutored opinion, crazy, [but] it is a strain of thinking that can be found at the highest level of the tech world.

Before I say much more about the movement itself and its vision of the future of humanity, I want to say a little bit first about how I came to this topic and why I was drawn to it. I don't think of myself as a science writer. In fact, I'm barely scientifically literate, as will probably become more and more apparent as the evening goes on. I'm not going to pretend that that wasn't an issue when I was writing the book, but hopefully not as much as you'd imagine. I have a PhD in English literature and my background is as a literary critic. I've always been especially interested in art and mythology that tries to get to the root of the strangeness of being human, in particular the existential kind of motherload of anxiety about our mortality.

I've been obsessed for years with the biblical story of the fall, which for me is an incredibly rich and profound poetic account of our own strangeness to ourselves, our own inability to accept ourselves as the animals that we are. The psychological and sort of mythic center of this story for me is the notion that we weren't supposed to be this way. That we were meant to be exempt from suffering and death and human frailty, and that we brought this condition on ourselves as a punishment.

Fundamentally, what's always been most fascinating to me is the idea of our own nature as something to be transcended. This inability to reconcile ourselves to the irreducible facts of our humanity has always been, in some sense, a defining aspect of the human condition.

When my son was born, five years ago now, I found myself thinking about this stuff quite a lot. The experience of watching my wife give birth and of being partly responsible for the existence and continued well-being of this tiny fragile little human led to a period of obsession with mortality and fragility. I don't know whether this is just me, but whenever I have a sort of a major life event, I always wind up getting obsessed with death; probably it's just me. It also happened when I got married; sort of like, this is the point now; after this point, where now commences the time [when] I die. It also happened with having a child. Probably just me.

It was around this point that I started to read a lot about transhumanism and to get quite obsessed with the idea of this movement that sort of explicitly offered a way out of this condition of mortality, a way of transcending our sort of human nature through technology. I was initially so fascinated by it because it seemed to me to arise out of the same basic existential unease with the human condition that lay beneath the myth of the fall and beneath my own increasing preoccupation with these things.

Although it took me to some very strange places and encounters with ideas [that] I found very troubling and very alien, the book itself had its genesis in an irreducible core of identification, of sympathy. I didn't like where transhumanists wanted to take us necessarily, but I totally got where they were coming from.

In retrospect, I think what was going on was what a psychoanalyst might refer to as a process of sublimation. I was channeling my anxieties about mortality and the fears of early parenthood into this obsession with transhumanism. And this obsession turned into a project and the project turned into a book, which is, I think, one of the few undeniably sweet deals about being a writer. You can generally find some way of turning your anxieties into something productive.

Transhumanism as a topic also appealed to me for another, perhaps related reason. It seemed to me to be a kind of extreme metaphor, a term I've borrowed from J.G. Ballard, who was very fond of such things. An extreme metaphor for capitalism, for the inexorable force by which capitalism blurs the boundary between humans and machines, applying a remorselessly instrumentalist logic to human life. In its conjuring of a future in which technology will allow humans to live forever, to merge with machines in a way that amounts to an evolutionary leap outside the confines of animal nature, any consideration of transhumanism inevitably leads to one quite troubling question: Which human beings? Who will benefit from these marvelous technological innovations we can expect? Will it be you or me or will it be the 0.01 percent, people like Peter Thiel and Elon Musk? What kind of world would that be? A world in which the super-rich transcend their humanity and leave the rest of us to founder in this state of biological impoverishment.

You can see perhaps how I came to view this movement as speaking directly to anxieties about the way our world has already been going for a while now, as a kind of fun house mirror providing a grotesque intensification of the way things are. It connected me with anxieties I had about the presence of technology in my own life, about the extent to which so much of how I lived, the decisions I made, and the things I consumed [were] already mediated by unseen algorithms.

And so, transhumanism was a way of thinking about all these things. As a topic, it was much less about forecasting the future than it was about getting beneath the skin of the present. If that sounds pretentious, that's because I am at heart extremely pretentious.

Many of us console ourselves with the thought that, as Philip Larkin's much quoted line puts it, "What will survive of us is love." Transhumanism offers us something very different and much less abstract, in a way. What will survive of us, in this view, is data. What will survive of us is code. It offers us a vision of the future in which our minds will be translated into zeros and ones and transferred from the flesh and blood of our bodies to an entirely different platform uploaded to machines.

The concept of mind uploading is a central article of faith among transhumanists. It's the key to the whole post-human immortal future. One of the most prominent proponents of this idea is the futurist Ray Kurzweil. Kurzweil is the inventor of a lot of ingenious devices, including the flatbed scanner, and he's the founder of Kurzweil [Music Systems] with Stevie Wonder. A few years back he was brought in by the Google directors as director of engineering. He's probably most famous for a book he published in 2005, which some of you might know, called *The Singularity Is Near*, in which he outlines how in the very near future—he puts a date on it, he says 2045—artificial intelligence is going to become so sophisticated and so powerful that we'll be able to upload our minds to AI supercomputers and merge with this technology in a final liberation from biology itself.

This, basically, is the notion of the singularity, which is often referred to more or less dismissively, but not entirely unfairly, as the Rapture of the Nerds. The picture Kurzweil paints of the future is one in which technology continues to get smaller and more powerful until such time as its accelerating evolution becomes the primary agent of our own evolution as a species. We'll no longer carry computers around with us, he insists, but rather take them into our bodies, into our brains, and into our bloodstreams, thereby changing the nature of the human experience.

Kurzweil's vision of the future is, I think, primarily attractive to people who already think of themselves as machines. People who agree with the artificial intelligence pioneer Marvin Minsky that the human brain happens to be a meat machine. Why would we, machines that we ultimately are in this view, not choose to upgrade ourselves to a higher degree of functionality? If we understand the machine to be an apparatus constructed for the performance of a particular task, then our task as machines is surely to think, to compute at the highest level possible.

In this instrumentalist view of human life, it's more or less our duty, pretty much the whole point of existing in the first place, to increase our computational firepower and to ensure that as machines we run as efficiently as possible for as long as possible.

Here's Ray Kurzweil in *The Singularity Is Near*: "Our version 1.0 biological bodies are frail and subject to a myriad of failure modes, not to mention the cumbersome maintenance rituals they require. While human intelligence is sometimes capable of soaring in its creativity and expressiveness, much human thought is derivative, petty, and circumscribed. The singularity will allow us to transcend these limitations of our biological bodies and brains. We will gain power over our fates. Our mortality will be in our own hands. We will be able to live as long as we want, a subtly different statement from saying we will live forever. We will fully understand human thinking and we'll vastly extend and expand its reach. By the end of this century, the non-biological portion of our intelligence will be trillions of times more powerful than unaided human intelligence."[1]

What Kurzweil is saying here, I think, is a version of what millenarian prophets have been saying for hundreds of years, that we will finally escape the fallen condition of our humanity and become un-fleshed. We will be restored to a prelapsarian state of wholeness, a final union in which technology will take the place of God. It's an apocalyptic sort of vision, in other words.

"The singularity," he writes, "will represent the culmination of the merger of our biological thinking and existence with our technology, resulting in a world that is still human but that transcends our biological roots. There will be no distinction," he says, "post-singularity between human and machine or between physical and virtual reality."[2]

Kurzweil has thought about the idea that such a merger would obliterate our humanity, and answers it by saying that it would, on the contrary, be an ultimate vindication of the very quality that's always defined and distinguished humans as a species, which is to say our constant yearning for a transcendence of our physical and mental limitations.

Transhumanism in this sense is in keeping with a strain in Western culture of thinking against the body itself, against the flesh, that is evident in much of Christian theology. In *City of God*, St. Augustine asks his readers to imagine a state of what he calls universal knowledge, far beyond anything we can now imagine, which will be granted to those blessed by the grace of God. "Think how great," writes Augustine, "how beautiful, how certain, how unerring, how easily acquired this knowledge will be and what a body too we shall have. A body utterly subject to our spirit, and one so kept alive by Spirit that there will be no need of any other food."[3]

One of the more radical transhumanists that I spent time with when I was writing the book was a software engineer from Pittsburgh named Tim Cannon. Tim was the leader of a group of what are called bio-hackers, a group called Grindhouse Wetware, in Pittsburgh, and they basically operated out of Tim's basement. They sort of designed and built technologies for implanting in their bodies, which gave them, in theory, sort of superhuman capabilities, but in practice very mildly superhuman capabilities, like the ability to sort of sense magnetic north or open electric doors by waving their hands. Which is kind of impressive, but only that, I think.

One of the guys in the company had just recently started work, so he had an RFID chip implanted in his hand, [but] because he was only a recent employee, he didn't have security clearance at the hacker space that they worked in, so the chip didn't actually do anything. Which I thought was quite poignant.

Tim had an interesting perspective on one of the main kinds of philosophical questions around all of this, which is a question that Ed alluded to a couple of times, which is to do with uploading—would it be You? What exactly would the uploaded simulated mind be? Tim told me that every eight years the human body replaces and regenerates all of its cells, so that you're literally not the same person as you were eight years ago anyway. You're basically like a rock band that has none of its original members, but that keeps touring under the same name.

Tim's point was that ontologically speaking, uploading your mind to a machine wouldn't be any different. Although I did look into this after I spoke to him, and apparently it's not quite true. There's a part of your brain, the cerebral cortex, where the cells remain throughout your life. So you're not quite touring without any original members. You're like Fleetwood Mac and your cerebral cortex is Mick Fleetwood. Ed, you can correct me on the science here, which may be a little off, or my pop culture references.

In one of my conversations with Tim, he said something that I felt was quite moving and very revealing of the religious subtext of transhumanism. He'd spoken quite a bit about his own past struggle with alcoholism and his problems controlling his own anger, his own sort of animal urges, and he was talking about how he felt hopelessly limited by the primitive and outmoded technology of his own human body. He said, "I'm trapped here. I'm trapped in this body." He didn't want to be limited in his thinking and his physical possibilities by this animal body that had evolved for life on the African Savannah hundreds of thousands of years ago. He wanted to be something more than human, more than the limited mechanism of his flesh and bone.

I told him this sounded to me like a profoundly religious sentiment, very much like something you might have heard from the Gnostics, an early Christian heretical sect who believed that we were divine spirits trapped in the flesh and that the flesh was evil. Tim didn't have much truck with this. Most transhumanists don't have much truck with any comparison between their ideas and religion. What they tend to say is that yes, there is some common ground between religion and transhumanism, in that both promise a final transcendence of the body, a final exemption from death. But the difference, they say, is that technology can ultimately deliver on these promises in a way that religion never could. Not only is transhumanism not a religion, they argue, but it is in a sense the end of religion, in that it offers us the prospect of becoming gods in our own right. This is the essential thing, I think.

I'll just finish with this, which is to say that eventually I began to understand transhumanism as a kind of paradoxical event horizon where rationalism, pushed to its most radical extremes, disappears into the dark matter of faith. I came to see it as a kind of magical rationalism, whereby technology takes the place of God.

M GLEISER: I think we have some things to think about. I'm sure you know that this year marks the two hundredth anniversary of *Frankenstein*'s publication, which is really a remarkable coincidence. I consider *Frankenstein* to be the first science fiction gothic novel and reading it, you soon discover that Mary Wollstonecraft Shelley knew what was going on in science at the time. And the cutting-edge science at the time was galvanism—that you could pass electric currents through dead tissue and it would move; it would respond with a jerky motion.

That was an amazing discovery, which inevitably led to the notion that, well, maybe the secret of life is electricity. If you're dead, Viktor Frankenstein thought, passing enough electric current through your body could bring you back to life again. Science would turn man into God. Unfortunately, it's not that easy, although something like it kind of happens when you apply a defibrillator to an almost dead person. Anyway, my point is that *Frankenstein* is a cautionary tale about the uses and abuses of the cutting-edge science of the time. It is no coincidence that Mary Shelley chose the subtitle *Or the Modern Prometheus*. We all know what happened to that poor Titan. I think we're sort of there again, two hundred years afterward. Science turning, or trying to turn, man into God.

This also reminds me of an *X-Files* episode called "The Great Mutato." It's about a mad scientist who is able to bioengineer different kinds of creatures by mixing different genetic codes, effectively creating a race of monsters. At some point, Fox Mulder, the FBI agent, breaks into the laboratory and asks him, "Why are you doing this? Don't you see what you're creating? Why are you doing this?" The immediate answer from the scientist was, "Because I can."

This is a very important point, which Ed ended with—talking about the need for getting together and talking about where we're going with this research and the potential dangers of blindly pursuing it. And Mark, you asked who is going to benefit from these technologies? I think that's an absolutely fundamental point.

My first question is: Agreed, we must talk about this. But who do we mean by this "we"? Who is going to make the decisions? Who's going to decide what's the right thing to do and what's the wrong thing to do? How are we going to figure out who is going to decide? Is it going to be a mix of politicians, scientists, philosophers, theologians, the general public, everybody? And how do we establish an equitable decision process that is going to guarantee what sort of safeguards could be put in place and implemented in practice, and that different points of view are represented?

E BOYDEN: I think in medical ethics there are principles—compassionate use, do no harm and so forth—which have been great at guiding thinking, and one of the directions that I think could be explored is this idea of the Asilomar of the brain. Maybe it's time for us to bring together different stakeholders—religious leaders and lawyers and scientists and doctors and so forth—because it is a kind of decision that affects everybody.

Questions of fairness should be discussed, as well. What kinds of access to technologies will be allowed by aspects like regulation or price? I think having that discussion is an important thing and I do take hope from this precedent over the last half century or so where at the dawn of the genetic engineering age, people decided that they were going to move forward, but in a way that allowed for transparency and openness and a discussion to occur.

M O'CONNELL: I think it's an absolutely necessary conversation. One of the things that I realized incrementally when I was writing the book, specifically when it comes to technology and the sort of technology that has infiltrated all of our lives in the last fifteen to twenty years or so, is that the extent to which the future or the present that we live in is not the product of large democratic conversations, but really the creation of quite a small number of very driven, very intelligent, very wealthy (in most cases) people.

I'm speaking largely about the consumer technologies. The present that we live in, none of us chose that, I think it's fair to say. We might choose it at a micro level. We might choose to log on to Facebook or to sign up for whatever, but we didn't have any part in the design of this technological present that we live in. Is there something to be learned from that, about these much more radical technologies? I would hope so.

M GLEISER: Here's another point. As scientists, it's hard to get enough grants to fund research, so we are partnering more and more with companies and industries that will fund our research. And of course, these companies have boards, and they have stockholders. They have interests that may be in conflict with the moral and ethical choices that we can and should be making.

This is going to be a real problem. A lot of people will want to use these technologies to make money and not to save the world. It's already happening, as we can see with the big tech companies. What may have started as a dream by a few tech visionaries has been transformed into a vast money-making machine that has changed the world we live in. This is an issue that has to be on the table from the start.

E BOYDEN: Thinking of the nature of ethical decision making for technology: you might make a decision [that] has positive effects at a certain point in time in the future but has negative effects at a far later point in time. Thinking about the medical world: at least for the brain, almost every therapeutic on the market has some kind of side effect. Or as another example, the technology that allows huge amounts of energy to be devoted to human production of goods is now resulting in climate change.

I think one of the key questions is really to consider the impact of technologies at different points in the future and have a view, not just of the shortest term, which maybe the profit motive encourages, but what does the human species look like ten, fifty, maybe even a hundred years or more after.

M O'CONNELL: One of the first experiences I had with the transhumanist movement when I was writing the book was [when] I went to an event in London, a meeting of a group called the London Futurists. The topic of the discussion was the ethics of cognitive enhancement, particularly around the question of distribution of this kind of future enhancement. Who would get the enhancements, and should there be a sort of socialistic kind of process put in place where, if there are enhancements to be had, everyone gets them, or should it just be market dictated?

Although I'm skeptical about some of the broader pictures of the transhumanist future, of this uploaded, upgraded humanity, it strikes me that if it were to become reality, it seems likely that these technologies will almost inevitably fall into the hands, first and foremost and maybe absolutely, of just those who can afford them. If the pharmaceutical industry is anything to go by, the profit motive is going to be primary.

I read a profile of Peter Thiel in the *New Yorker* from 2012 or 2013, and one of the major topics is his investment in the idea of radical life extension. And the journalist who's profiling him asks him basically this question. He says, "Wouldn't the technologies you're talking about and funding, if they were to become a social reality, would they not just wind up radically exacerbating the kinds of structural inequalities that we already have?" And Thiel's answer struck me as the most extraordinary thing I've ever heard, really, an extraordinary illustration of the kind of mindset that you encounter in this movement. He said, "To me, the most extreme form of inequality that exists is inequality between people who are alive and people who are dead."[4]

M GLEISER: Wow.

Alright, what if we bring this conversation closer to something that is happening right now? For example, self-driving cars and trucks and that sort of technology, which involves machine learning and artificial intelligence to a very high degree. It's happening and fast. I'm sure you all know that there was an accident in Tempe, Arizona recently where a self-driving car hit and killed a pedestrian. Of course, every cutting-edge technology, coldly speaking, is going to have failures, because that's how we make progress in science. It's through mistakes that we learn how to advance. But that's not what I wanted to talk about. What I wanted to talk about is job displacement, the transformation of the marketplace.

I [wrote] an essay recently where I mentioned that in the United States, there are about 1.5 million truck drivers and about five hundred thousand school bus drivers. Technologies like this, that depend on machine learning and artificial intelligence, will make many jobs [that] are how millions earn a living obsolete. What will these people do when jobless? And what is the right thing for the corporations that are developing these technologies to do? These technologies will create a huge number of displaced and aimless professionals, many of them too old to be retrained in the new jobs that are emerging. Shouldn't there be a call for corporate ethics here?

M O'CONNELL: I would be quite skeptical about the prospect of the people and the corporations that are driving the development of these automation technologies taking seriously the ethical implications. I think part of the logic of capitalism is to reduce overheads from labor, and obviously the most effective way to do this is to automate labor out of existence. Automation is capitalism on steroids.

The obvious problem there is the instability of that systemically. How can you have a system that continues to sustain itself when people don't have jobs and they can't afford it?

Interestingly, that's a conversation that has been had. As far as I can see, the really serious conversations are happening actually in Silicon Valley. You have traditionally quite liberal or libertarian people bringing up ideas around universal basic income and so on. The really serious conversations seem to be happening around that. I think a lot of it comes from an anxiety of well, what are we facing? Are we facing mass social unrest, mass unemployment? How do we live in those kinds of futures?

Again, it loops back to the issue of what we are building. Should we be having the discussion now whether we should even be building these technologies? It seems that there's a kind of inbuilt logic that if we can, it's going to happen. There's money to be made of course.

M GLEISER: Do AI scientists talk about these things at all? Do you have conversations like this with people in your lab, or just with people like Mark or perhaps some scientists, like me, who actually worry about these things? I'm very curious. I really don't know because I'm not part of this whole movement. What I do is very different—cosmology and the origin of the universe. Fortunately, we're not going be making artificial universes any time soon. AUs are for the distant future perhaps.

E BOYDEN: Yes, it's a conversation that I try to have a lot. I think that especially the jolt in the arm from technology entering neuroscience is accelerating the pace of such conversations. When President Obama launched the BRAIN Initiative, which was all about building new technologies for mapping and repairing the brain, having an ethics commission to consider the ethics of a given human experiment was one of the top priorities. So, I think that it's happening.

What I am hoping for, and that's why I alluded to it in the presentation, is to broaden the conversation to include more stakeholders, to include companies. If companies feel like they're left out of the conversation, maybe they will think it's not a legitimate one. But if we can make them part of the conversation, in the best case, maybe better solutions that people will comply with will emerge. And in the worst case, it would still demand some kind of accountability from those present, which is not bad. Are there ways that we can make this something that people want to talk about?

Part of me worries that people don't like to talk about the brain. I was on a panel once at a conference with a bunch of business leaders and the person chairing the panel asked the audience how many people in the audience had taken a cognitive enhancing drug. Nobody raised their hand. Then he revealed his secret poll he had done earlier: it should have been around 20 percent. People didn't want to talk about it. I worry that in our current language about the brain, people feel it's dehumanizing to think oh, my brain is generating this action. It's sort of removing our agency from our identity. Sometimes I feel like we need a new word that means, all at the same time, "this is me and this is what my brain is doing," and allows us to avoid the dehumanization of current terminology, so that the brain becomes something people want to talk about. I think that's an unsolved problem, but maybe one that could be addressed linguistically.

M GLEISER: You could raise the level of drama a little bit, perhaps. You could say something like, with nuclear technology, for the first time in history, humanity has the tragically dubious power to obliterate itself. Now, with biotechnology plus AI, for the first time in history, humanity has the power to reinvent itself. That's a big deal because we'll just become something else. We don't quite know exactly what that something else is, and that's scary. As you very well said—and I'm really happy that you were humble enough to say that there's too much information about the brain we don't know how to get—there is no way we can actually capture the whole thing. Whatever we end up simulating or creating, it won't be a human brain in all its details. However, we're going to do what we can to create smaller brain-like simulations and then, of course, make them bigger and bigger, even if the complexity scales in a highly non-linear way. It's impossible to predict where this will lead us.

E BOYDEN: There's another way to look at it too. That's why during the little mini talks we gave, I tried to bring up the idea of enlightenment before the idea of augmentation. That has both humanistic as well as practical implications. On the practical side, augmenting what one doesn't understand can backfire. For example, the neural implants that are helping people with Parkinson's disease, they're doing a lot of [good], but we don't understand the circuits in the brain that are being perturbed. So simulation can also result in changes in emotion or cognition that we can't currently predict.

I was a bit alarmed with the DIY brain stimulation community, where people apply currents to their heads to try to boost, for example, video game performance. To be honest, I don't know if anybody's done long term studies of what happens, over periods of months to years, to learning or to thinking or to creativity. In part, that's because these technologies just haven't been around that long. But part of it is also, how do we want to think about the brain? If we really want to treat the brain as an engineerable thing, what are the principles of such engineering? If you're building a computer or building a machine like a car, there are certain principles of design where you do something like this and not like that because you know what could go wrong. We don't yet have that knowledge for the brain because the science is so young.

M O'CONNELL: One of the things Ed's talk got me thinking about was the issue of the sort of blurry line between enhancing basic human functionality and more therapeutic applications. It seems to me that when we use terms like transhumanism, often it's not clear where the dividing line is between the two. There's a way in which you can describe or define transhumanism as just the imperative to use science to improve the human condition, which is really what science is, or at least what medical science is, certainly. So, it strikes me that there's a lot of the stuff that you were talking about within the realm of traditional ameliorative kind of applications of technology, but very close to that border as well.

E BOYDEN: Maybe there are differences of degree. There's a certain pace of human change, which is always ongoing. If you were to go back a hundred years, of course, it's obviously a world without computers and airplanes and all sorts of stuff. Maybe today would look like a transhumanist world to somebody a hundred years back. From your talk, one of the things I was thinking about was maybe the mismatch between what one hopes for and what one can see a realistic path to achieving. Do you think that gap might be what leads to some of these religious perceptions?

M O'CONNELL: I think so. I think maybe that's where transhumanism falls, actually. That gap is another way of describing the difference between what you were talking about, Ed, and what transhumanists are talking about.

E BOYDEN: When a gap closes, then I think our tendency as humans is to take it for granted. Penicillin, for example, to pick maybe an extreme example. We have these invaders attacking the body and here is a pill you're taking to wipe out all the invaders. You can make it sound very dramatic. It seems almost magical when I describe it like that. Now it's something that billions of doses have been given out and countless lives saved. And it's become ordinary. Maybe it's because, again, our expectations have matched reality there.

M O'CONNELL: We are already post-human. Part of the definition of humanity is transcending humanity, in a way. Transhumanists are very fond of, and I think rightly, pointing out the way in which we're all already cyborgs. If you wear glasses or if you have a pacemaker or maybe some kind of attachment to your face that makes your voice louder or whatever, you're already in a kind of a cyborg relationship with technology. It's a very blurry kind of line that we're attempting to draw.

M GLEISER: To me, the smart phone is our cyborg extension. Think of when you forget your cell phone one day at home—it's desperation! Who are you without this machine? This machine is now you. You and this machine are one. In fact, if you go and you scan everybody's cell phones here and you look at the different apps, certainly a lot of them will overlap, but each one of your cell phones will be unique to you because it is in a sense a digital fingerprint of who you are. It's already happening, a point I made in a previous dialogue with Patricia Churchland and Jill Tarter. Cell phones are digital extensions of our carbon selves.

M O'CONNELL: I mean, if I were to reach across and take your phone, it would feel like a personal invasion of space.

M GLEISER: And now for some questions.

QUESTION ONE: Mark, you mentioned [two] things about singularity: the cerebral cortex cells not dying, and who chooses who gets these technologies if and when they become available. Do you see the cells in the cerebral cortex as a model of how that might get decided? It almost sounds like those cells are the immortal ones in the body.

M O'CONNELL: The ones that . . .

QUESTIONER ONE: That don't die.

M O'CONNELL: Remain.

QUESTIONER ONE: Yeah.

M O'CONNELL: Actually Ed, can you verify that is in fact true before we continue?

E BOYDEN: So brain cells do die. Most of them don't get replaced. There are some parts of the brain, this is actually turning out to be kind of controversial, that do regenerate brain cells throughout life. But if there's an injury, or over time with aging, then cortical cells do degrade.

I think maybe a different spin would be useful, is there another metaphor?

M O'CONNELL: It's difficult. My cerebral cortex is quite slow. Again, that's completely the wrong language to use. To get back to your question, you're asking whether the cerebral cortex is the thing that . . .?

QUESTIONER ONE: I guess I almost feel like that's something that kind of falls into place. If we were to use the brain cells as a model, I don't think anyone ever or any cell ever decided hey, these guys can go live on forever in the cerebral cortex and everyone else can regenerate every seven days.

M O'CONNELL: OK. So you're saying at species level the people who . . .

QUESTIONER ONE: Right. On a cellular level. I would say the analogy would be the human body being society and us being the cells in society.

M O'CONNELL: That is an interesting analogy. Is it slightly Darwinian in a sense that as long as a certain small subsection of us continue on and evolve then that's an evolution for the species? I think it's confusing at this point.

M GLEISER: Let's move on. We want to have a lot of people asking questions if we can.

QUESTION TWO: In a moment of serendipity, I actually read an article earlier this week that handled transhumanism. It was about a book by Alan Jasanoff at MIT called *The Biological Mind*, where he challenged the idea that we'd ever be able to download the brain to hardware. He was saying that the brain is a much more complicated part of a biological system, where portions of our consciousness are housed in other parts of the body. Ed specifically brought up the idea of how our brains are influenced by our micro fauna, so I wanted to ask if you've thought of anything about that. If you think that brings up any challenges for this transhumanist idea. How that affects—how this changes our identity as humans.

E BOYDEN: [Alan Jasanoff] and I co-direct MIT's Center for Neurobiological Engineering so we talk about these things every now and then. I definitely agree. We now know that, as you noted, gut microbes can influence the brain. The brain and the immune system talk. What we eat, drinking a cup of coffee—obviously our environment affects us. This is the concept of embodied cognition. This idea that you have to treat thinking, feeling, and consciousness as systems. I think that is pretty clearly the case.

I think in the terms of the scientific agenda, to understand one could extend it beyond the brain. In fact, in our research group, we're now actually thinking about mapping the immune system and the rest of the body. We're thinking about how you integrate the sensory environment, and what your movements do upon the world, into pictures of the brain.

From a strict scientific sense, I think that the technologies we build for mapping the brain could be used for other bodily systems as well. But it doesn't trivialize the complexity. These are extremely complicated things but maybe it's not as bad as one would think because the tools are getting better and better.

QUESTION THREE: Both of you talk about the consequences of technology after we get it. One of the things that I'm curious about is this idea of who decides what needs a cure. Who decides what's a problem?

One of the things that I was thinking a lot about during both of your talks was this idea of dialogue—like "solving" LGBTQ people. That's something that I don't think is a problem that needs to be solved. This was something that I saw when I studied American Sign Language. One of the big distinctions when you're learning American Sign Language is the difference between somebody who identifies as Deaf with a capital D and somebody who identifies as deaf with a lower case. The capital D is usually somebody who was born deaf, raised in the deaf community, and has experienced an entire culture that hearing people don't get to experience. One of the things that comes out in their community a lot is this discussion of cochlear implants, for example, destroying Deaf culture. So I'm curious, who decides what is a problem or what needs to be solved to begin with and are those conversations happening even before we start these technological studies and corrections.

E BOYDEN: I think, again, the discussions need to be broadened to include not just a doctor, not just a scientist, but the broader community. At one point, I was diagnosed with Attention Deficit Disorder and put on Ritalin. After a year of it, I realized, hey, I'm just going to deal with it. Maybe it's a different way of approaching problems and maybe I'm good at certain things and not so good at other things and that's just going to be the way it is. A long time after that episode, which was when I was a teenager, it made me think a lot about who decides what gets a label as a disease state?

I think part of the goal is to facilitate mutual education: "Look, this is how I feel and how I think. Oh, but here's other things that I need to learn as well." If that mutual education process can occur in a non-threatening way, my hope is that we can make those discussions happen. Again, I do feel it's the kind of thing that's uncomfortable to talk about. In fact, I [have] rarely talked publicly about my own experiences with Ritalin because it can seem a little bit weird or stigmatized. But maybe by talking more about it we can make it a discussion people want to have.

M O'CONNELL: I think that's a really interesting question and touches on, for me, one of those pressure points of the whole ideology of transhumanism.

It makes me think of one of the people that I spent time with for the book. He was a guy called Zoltan Istvan, who ran for president in 2016. He formed a group called the Transhumanist Party, and I went on the campaign trail with him. Zoltan was one of these people who would put out any kind of opinion just to put ideas out there.

When I was with him, he had just published a very controversial opinion piece in, I think, psychology.com or *Psychology Today*. He's from L.A. and the city of L.A. was earmarking x million dollars to make downtown L.A. much more wheelchair accessible. Zoltan wrote this article from a transhumanist perspective, basically saying that it made no sense to make L.A. wheelchair accessible. What needed to happen was that the government needed to earmark funds to upgrade people, like wheelchair users.

This was obviously hugely controversial and outraged, rightly so, a lot of people. Zoltan didn't quite get why people were so offended by it because, as with a lot of transhumanists, his thinking was well, we're all disabled by virtue of being in these human bodies, so we all need to upgrade ourselves.

I think that's one of the things that, for me, really hits on this quite visceral rejection button in me. I'm not quite sure where that sort of morality comes from, but that's one of the points where I break most firmly with the transhumanists that I'm writing about. This idea that we need to upgrade ourselves because we're somehow insufficient. That's troubling to me and I think it's troubling to most people.

QUESTION 4: Are you familiar with the term connectome harmonics? Well, I'm not too familiar with it. The gist of my completely non-scientific brain's understanding of it is the idea that the brain wave activity and patterns that happen, the forty hertz going on in synchrony with also eight hertz in synchrony with eighty hertz forms—I'm a professor of music so I'll continue music analogy, let's say, a piece of music. I'm reflecting on the rock band analogy. One of the things that struck me as odd about it was, it seems as if you lose the essence of the band when

Mick Jagger walks away and Jick Magger walks on stage, yeah. But the music they're playing is still the same and you don't think the same thing when you think of the symphony. When you think of the Boston Symphony Orchestra, the players are constantly cycling in and out, but the music stays the same.

When I think of definition of consciousness—who we are, our identity—I think of what decisions would this person make. By that definition, I am not the same person now as I was twenty seconds ago when the microphone was handed to me.

What do you think about that?

M O'CONNELL: That's beautiful. Well, I'm immediately thinking maybe it's actually just another question. Are you the symphony orchestra or are you the music or are you some strange combination of both? I think that's a brilliant shift of analogy actually from the rock band to the symphony orchestra.

M GLEISER: Don't forget the conductor. The Boston Symphony plays with different conductors; so, Beethoven's Fifth is going to be different even though it's the same music on the score. It's going to sound different. There is an imprint there [that] is original and unique.

E BOYDEN: I do think this idea of identity over time is really interesting. In some ways, I'm very different now than ten years ago and I'm subtly different now than five seconds ago. Defining what that is, I think, is something that we can't quite yet do very well in neuroscience. Maybe with new kinds of brain imaging technologies or brain mapping technologies, one could try to define the differences or similarities and see what they are. I wonder if it might be possible at some point.

QUESTION FIVE: In religion and spirituality there is this concept of soul that can live on forever. I'm curious if in transhumanism [there] is anything equivalent to that or is it just this idea or concept of uploading your consciousness into a computer?

M O'CONNELL: One of the things that I realized very quickly about transhumanism was that I was encountering a group of people who were quite hardline rationalists and materialists, basically. One of the interesting contradictions about particularly the idea of mind uploading is [that] although it comes from a very hardline materialist kind of point of view, like everything is physical, everything is firing neurons and so on, there's also a kind of—what's the word, the opposite of materialism? It's gone.

E BOYDEN: Dualism?

M O'CONNELL: Dualism. Yeah. There's a kind of dualism inherent in it as well. There's the mind as this kind of immaterial thing that can be extracted from the brain. There's a kind of a contradiction there. I guess the soul is possibly just another way of talking about mind, just an older more religious way of talking about it potentially. But no, in general, as I said in the brief talk, transhumanist tend to not have any kind of truck with religion. Although I did attend a conference on religion and trans-humanism in the Bay Area, which was very poorly attended because it was boycotted by most transhumanists. There were some interesting crossovers—there were Buddhist transhumanists and a Muslim trans-humanist and a lot of Mormon transhumanists, for some reason that I never quite got to the bottom of.

NOTES

1. Ray Kurzweil, *The Singularity Is Near: When Humans Transcend Biology* (New York: Penguin, 2005), 9.
2. Kurzweil, 9.
3. Saint Augustine (Augustine of Hippo), *City of God*, trans. Marcus Dods (Digireads.com, 2017), 705.
4. George Packer, "No Death, No Taxes," *New Yorker*, November 20, 2011.

7

ON HUMAN AND PLANETARY LONGEVITY

A Dialogue Between an Environmentalist and a Doctor

MGLEISER: This conversation is on longevity, human and planetary, because obviously you can't think of one without the other.

I thought it would be very interesting to have a conversation on the different scales we use to frame the problem. We can, of course, think about it from a human perspective. We humans face this very perplexing dilemma—we are self-aware creatures [who] know that our life on this planet is limited, bracketed by our birth and our death; that is, we are animals aware of the passage of time, a strange superposition of two clashing natures. On the one hand, we are animals. We have hair and nails that grow. We reproduce like most mammals do. We feel physical pain and hunger. We are deeply dependent on the natural environment. On the other, we contemplate the sublime. We ponder the infinite. We can think about transcendence. We think about change and transformation. And we know we are alive, and we have an understanding of what that means.

In 1973, the cultural anthropologist Ernest Becker wrote a book, which won the Pulitzer Prize in 1974, called *Denial of Death*. In this book, he explores precisely this issue. How do we find meaning in life, with these two extremes pulling us in different directions, the animal coexisting with the demigod? It's a very complicated conversation, and one not without existential angst.

At about the same time Becker published his book, in the 1970s, the first models of climate change were being developed. These studies brought out into the open the realization that first, humans impact the weather, and second, that this impact could have devastating consequences to our collective future, representing a serious existential risk to our species and countless others. Tonight, we will explore mortality, the extension of life, and extinction from a human to a planetary scale.

Elizabeth Kolbert has been a staff writer at the *New Yorker* since 1999. Her three-part series on global warming, "The Climate of Man," won the 2006 National Magazine Award for Public Interest, among other honors. She received a Lannan Literary Fellowship in 2006, and a Heinz Award in 2010, and won the 2010 National Magazine Award for Reviews and Criticism. She is the author of *The Prophet of Love: And Other Tales of Power and Deceit*, *Field Notes from a Catastrophe*, *The Sixth Extinction*, for which she won the 2015 Pulitzer Prize for General Nonfiction, and most recently, *Under a White Sky: The Nature of the Future*.

Siddhartha Mukherjee is a pioneering physician, oncologist, and author. A very influential voice in the scientific community, he's best known for his books *The Emperor of All Maladies: A Biography of Cancer*, which earned him the 2011 Pulitzer Prize, and *The Gene: An Intimate History*, which was recognized by the *Washington Post* and the *New York Times* as one of the most influential books of 2016. *The Emperor of All Maladies* was adapted into a documentary by Ken Burns and included among *Time* magazine's [All-Time] 100 Best Nonfiction Books.

S MUKHERJEE: I thought I would begin with a thought experiment. And the thought experiment goes like this: If you were sitting in the 1850s, and you asked the questions, "What is it like to get old? What is it like to live beyond fifty years old?" someone might say to you, "It's a horrible life. Your teeth fall out. Your hair degenerates. Your skin degenerates. It's a terrible thing. And what we should really do is respect human boundaries and dignity, and let people die quietly at fifty."

Fast forward a few years, and all of a sudden, those realities—your skin degenerating, your hair degenerating, your body, you mind falling apart—start to change. Because all of a sudden, we begin to make important advances, scientific advances, cultural advances, advances in public health, hygiene. When all of a sudden, fifty isn't a boundary. And that same person who told you, "Why not just give up at fifty? It's a terrible life out there," would start saying, "Well, actually, now sixty sounds good." And, so on and so forth. Sixties become seventies.

So, there are two strands in our understanding of longevity of the body. One strand constantly reminding us that death and aging and the end is nothing but inevitable, and that's, of course, true, but on the other hand, another strand pushing and saying, "Well, what's inevitable about it? What is it about . . . certain organisms live much longer. What are the biological limits?" Well, the biological limits are defined by what we want to understand and how we want to push them.

So, I think that there is—as a physician, as an oncologist, as someone who works in cancer, you feel both these strands. And it's important to remember that the two strands, it is absolutely true, it's absolutely true, that there is a kind of collapse of dignity when you begin to say, "Let's all live 'til one hundred." And, that changes, it distorts certain ways of thinking about culture. But, on the other hand, let's not forget that, as I said, a hundred years ago people would say that living beyond fifty was a horrible life.

With that as a background, I thought I would propose to you three potential visions of longevity or immortality. They're quite different from each other, and each one has a very different consequence about how we think about human future. The first one is the most traditional one. That's the one that we've grown up with, in a kind of cultural sense. It's immortality viewed in the negative, or in the double negative. That's simply, if you killed all the killers, you would achieve infinite longevity, or at least true longevity. This is the stuff that I came into my professional life, my scientific life, with; take diseases away. Remove cancer, heart disease and other such diseases, infectious diseases.

If you look at human history, it's an astonishing history in the last hundred years, because the conquest of communicable diseases, so-called infectious diseases, lengthened human lives, human longevity. Whether that conquest was through public health and hygiene measures, some vaccinations, use of antibiotics—a mixture of things—that changed the epidemiology of human disease. Most of us would be dead in this audience. We would be all long gone 100–120 years ago. Virtually none of us would be here, actually.

Then the epidemiology, the face of human disease, began to shift and become more and more focused on the so-called non-communicable diseases. So much so that it feels these days that the division of the world between non-communicable diseases and communicable diseases, infectious and non-infectious diseases, is a little bit like saying mammals can be divided into cows and non-cows. That's because the category of non-cows, non-communicable diseases—heart disease, cancer—even in countries that you wouldn't imagine, even in parts of the world where you would think that the burden of communicable diseases is enormous—like Tanzania, just to pick one out of a hat—the pattern is shifting so dramatically that there's an expectation that more people will die from the complications of hypertension than they will die from the various complications of infectious diseases in ten years. There's the sense of the changing nature of epidemiology.

If you killed all the killers, if you removed the constraints, or the barriers, to long life, you would actually increase human longevity enormously. And, of course, there's a truth to that.

Then came a second vision, which began to appear. And, it comes from a variety of sources, which I'll talk about. It turns out that if you look at people who have extreme longevity, people who live 120 years, 118, 108, 110, what's interesting about those people is that they are not killed by killers. They, in fact, are dying from diseases that we don't think of as diseases.

For instance, they die of peculiar forms of bacterial sepsis. Their immune systems collapse. They sometimes die of broken bones. My laboratory, actually, found stem cells that make bones. There's a project that we're doing right now to try to find people with extreme longevity to ask the question, "What happens to stem cells in their bone?" We discovered these stem cells some years ago. We're trying to put them back into human beings. They make bone and cartilage, among other things.

The second idea is that, sure, you can take away the killers, you can kill the killers, but there may be genetic, biological, epigenetic, or other changes that prospectively will allow us to extend human longevity. In other words, there are genes, or diets, or bionic mechanisms by which you could alter your lifespan in a positive way.

Imagine I would find a gene in a small animal that increases longevity enormously (and such genes have been found). Interestingly, many of them have to do with the way we regulate metabolism. With many of them there's a very careful balance between those genes and the capacity of them to tip into cancer. There's a very tricky balance, but occasionally, you can find genetic means of changing the limits of human longevity.

One interesting consequence that arises out of that is to say, "Well, forget about killing all the killers. Maybe we should be tampering with human genomes so that we can increase longevity." I would say, in trans, which means that they suppress all the effects. They would override any other effects and it would be non-disease-specific. They would essentially override the effects of diseases.

Do such genes exist? Yes. They exist in other animals. They exist in worms and in flies. You can make worms that live three times longer than other worms. Often, these genetic means cluster along pathways, along very particular ways. Often, they cluster on metabolic pathways. For a long time, you would imagine that these alterations—unless all of a sudden there appeared a genetic mutation, a genetic variation in humans that would spontaneously arrive and that could give that kind of longevity—you would imagine it would be very hard to tamper with this physiology. But certainly you can tamper with the physiology in other animals. Worms, as I said before.

One thing we'll talk a little bit about: with new genetic tools, it's possible to tamper with the human genome, in principle. Again, remember you can do it in two ways; you can remove constrainers, and we'll talk a little bit about that, but also you can change in a positive direction. You can not just remove the constraints, you can influence in a positive direction, just like in small animals.

Here the world splits into, generally speaking, two kinds of camps. One camp would suggest that longevity and all these complex traits are so complicated that there'll be many, many genes, and possibly across . . .

Just to give you a sense of what the scope of the problem is, if you were to take the human genome, and you were to print it out in standard print format, a single human genome would be 66 full volumes of the *Encyclopedia Britannica*. This room, wall to wall, all over, would be Elizabeth's genome. And in the past, it was largely inscrutable. It's written in just four letters. If you opened up one volume, you would read A-C-T-G-C-T-T-T-G-C (I'm making it up).

In the past, our capacity to read that information was extraordinarily limited. You could say, in volume 36, on page 72, I have a genetic sequence that encodes for the cystic fibrosis gene. And if you change this letter to that letter, you could go from a normal cystic fibrosis gene to an abnormal one. And if you have two copies of that, and so forth, then you could make some conclusions about it. These were single genetic changes [that] we could map onto human traits, diseases, and so forth.

Now we are getting [to] a place where we're actually beginning to read this information in a powerful manner. To give you a sense of how powerful that can be, and how disruptive that can be, I just read a paper two weeks ago, still in pre-print, where they took human height as a characteristic. Height's a very interesting characteristic because we know it's highly heritable. Tall parents tend to have tall kids; shorter parents tend to have shorter kids. This was known for a long time. But for decades, people have been trying to find a so-called gene for height, single genes for height. For most human variation they were very hard to find. And we knew that there were probably tens of thousands of variations. There are only twenty thousand genes, but the lettering is very wide. There were tens of thousands of human variations, each with small effect, that would explain height.

The problem, of course, was that in order to decipher that you would have to create a compendium of all human beings and all their heights, find all their genome sequences, and try to find a machine, a computer, or an algorithm that would correlate one with the other. All of these resources didn't exist, but it turns out, now they do exist. And this particular paper uses a powerful computational algorithm to predict human height from genetic information alone. This is assuming no malnourishment, a very important constraint, assuming no environmental change. From human genetic information alone.

Think about that for a second. That means that, in principle, you would be able to predict the height of your unborn fetus before that child was ever born. You would know what height they would attain. And now, think about extending this, potentially, to other attributes—color of hair, the color of your skin. All of a sudden, you've gone into territory [that] I think all of us would feel a little uncomfortable with.

Longevity rides along similar principles. We don't have fundamental notions about it, but it would be very, very clear that the heritable part of longevity, the part that can be inherited, will begin to be solved. In fact, that's a project going on right now. This is one of the great bio bank projects. Then we would be able to read longevity. If there are genes [that] very strongly influence longevity, or if there are genes that very strongly influence longevity in the negative direction, we can, potentially, change those.

There happens to be one already, a big candidate, a gene that increases your risk for neurological syndromes—Alzheimer's disease, neurodegenerative diseases—called APOE. And [people] like George Church, or others, have made a compelling argument. Why not take it away from the human genome? Why not remove that? Why live with this constraint if you don't have to live with this constraint? That would be the second vision of longevity, tipping into far edges of mortality. It's not immortality, but far edges.

Then, the last one is perhaps the most intriguing one. I've been thinking about this as a thought experiment for a long time and the question that I posed myself a while ago was, "Could there be a soft version of immortality? Could we hack it?" By that I mean not inventing crazy new tools, but what if someone gave me the toolkit of what's available? Could I make soft immortality possible? I began to think about the following thought experiment.

Let's say I used you, Elizabeth, as an example. We are quite confident that we could clone you. We can clone most mammals. It's not easy, but within limits I think the possibility of cloning is becoming more and more clear. And if it's not cloning, we could even do a soft version of cloning, in which we would derive sperm and eggs from your normal cells, combine them, and basically recreate a soft version of a clone. It wouldn't be the Dolly version of the clone, which is more complicated, in which you have to take another cell out, suck up the nucleus and re-inject it with your nucleus, thereby creating a clone. We could do it a bunch of different ways. We could recreate, basically, sperm and eggs, in principle, from someone's cells and then combine them again.

Let's assume for a second that we would be able to do the cloning piece of it. Then here's the experiment: what struck me was that the cost of digital recording of all your experiences is diminishing so dramatically that you could basically wear a GoPro, walk around, and you could edit your GoPro. You could edit your GoPro as you aged and you could say, "These memories were really significant for me. That time I went to Rio and sat on the beach. That moment I experienced that Glacier," etc.

My question was, what if we began to clone ourselves and give the clone a GoPro? Give them a quick GoPro lesson on your life, in which they would absorb and perhaps even experience what you thought were the fundamentally important pieces of you. You would have yourself in a second generation now absorbing pieces of you.

What's interesting about the experiment is that it's infinitely repeatable. Your clone makes a clone, gives a GoPro to the clone, etc., etc., etc. Of course, information is lost. It's not going to be your life, but I just keep wondering about this idea. It haunts me a lot. What if I [were to] do that? Would it alleviate some concerns that I have about my end? I would end, but very important pieces of me would be left over. A true genetic copy would be left over. And things I treasured, my most important memories, would be here.

It's dystopic, of course, but yesterday's dystopias are tomorrow's old news. There's a quality about this that makes you think. This is a thought experiment. I'm going stop there. That's kind of a broad introduction, but those were my three visions of longevity tipping into the edges of immortality.

E KOLBERT: Yes, speaking of dystopia. That was a perfect lead-in. I am supposed to represent the humanities here. I'm up here with two scientists, and the humanities always get the short end of the stick, the bum end of the stick. Sid talked about immortality, and I'm going to really flip that around. I think that there is a really interesting conversation—it's not a conversation, maybe, it's something much worse than that, a brawl—and that is to talk about what humans have meant to the rest of the species on the planet.

We don't know how many species exactly there are on the planet. Rough estimates are about 10 million. We have named maybe 1.5 million of them. There's a lot of life that we know from statistical tools is out there, but we really don't know what it is still, at this point, in 2018. We know that we're destroying it pretty rapidly. The reasons for this are the flip side of our tremendous success at staving off disease, at reproducing, at feeding ourselves. All the things that have brought us to the point where there are now 7.6 billion people on the planet, heading very rapidly to 8 billion in the next couple of years, actually. If you want to know exactly, there's a world population clock and you can watch that thing spin around, not just by the one or two or three, but by the ten thousands every few minutes as our population grows.

I wrote a book about this idea that we are causing a mass extinction. What is a mass extinction? A mass extinction is a very simple concept. It's considered a short period of time, relatively short on a geological scale, when a lot of the diversity of the planet, for whatever reason, crashes.

This idea that there's a sixth extinction obviously implies that there have been five previous ones. There have been, lo and behold. The first occurred 440 million years ago during a period known as the Ordovician period, when life was still largely confined to the water. About three quarters of all marine species died out, and very few on land, because there were very few on land.

The most familiar one of these mass extinctions is the fifth, which did in the dinosaurs sixty-six million years ago. There's a pretty broad scientific consensus now that that one was caused by an asteroid.

For a long time, people tried to find common themes in these extinctions. Is there one reason why they happened? They happened at irregular periods. No one has yet found the key, what causes a mass extinction. But the general theme of them is that the world changes in some way, faster than evolution can keep up with. It all does boil down to genetics.

If I'm a worm or a plant, I can only keep up with environmental change by—either, [one,] I have enough plasticity and I have enough adaptability. If I can't go here I go there; if I can't use this food source, I go to that food source. All organisms have a certain amount of plasticity. Or the other way I adapt is through mutations, through genetics, through evolution. I evolve. I become a new creature. And that's propelled life throughout the last 3.5 billion years or so.

If things change really fast, and I don't have the plasticity, and I can't evolve fast enough, the end is near. That's the end of the line for my species.

And what are we humans doing? We've sort of transcended evolution in a really, really important sense. This gets to a lot of the things that Sid was talking about. We don't wait around to evolve. If we have a new pathogen, for example when you go to get a flu shot, that flu evolved into a new virus, but you got the vaccine and you skipped that evolutionary step. The flu didn't just kill off everyone who wasn't immune to it. We jumped over that. And that's why we've done so well, why we've extended our own lives, why we're so successful.

But the flip side of that are these massive changes that we're making to the planet on a geological scale. I'm sure everyone in this room has heard of this concept that we're in a new geological epoch, the Anthropocene. An epoch is actually, to a geologist, a very small amount of time. There are many people who believe we're in a new geological period, a whole new period of time—a very big geological divide—because we're changing the planet on at such a rate, and on such a scale.

I'm going [to] throw out a couple of the ways we're doing that. Unfortunately, I could spend all night talking about how we're doing that, but I will not. I will turn this over to a conversation. We're changing the climate really fast. We're putting up CO_2 really, really fast into the atmosphere, perhaps as fast as has ever happened in Earth's history. And when you think about it, it makes sense. It's really hard to put CO_2 into the air. All the fossil fuels that we are burning are just carbon that a plant absorbed during its lifetime. The plant died, and for complicated geological reasons didn't decompose but became a fossil fuel. That process took hundreds of millions of years, and we are now digging up those stores of fossil fuels. We're burning them in a matter of a couple centuries. That's really, really rapid. And if people weren't here, how could that be happening?

We're acidifying the oceans. That same CO_2 that we're throwing up into the air that's changing the climate is changing the chemistry of the oceans.

We are very rapidly altering the surface of the earth. When we go into the Amazon, or when people came to the United States and mowed down the prairies and planted corn, that's a monoculture. That's a place where very, very few creatures can survive. Everyone's heard about how monarch butterfly populations are crashing. That's because we go into what used to be their habitat, we just mow it down, put down herbicides, plant corn, and there's just nothing for them to eat. It's as simple as that.

Other ways: we are moving things around the world. This one doesn't get as much attention as it probably should. Over the vast span of evolutionary time, the continents have divided. All the continents used to be squished together about 250 million years ago into a continent we call Pangaea. Then, the continent split apart, the world as we know it came into being. Evolutionary lineages have been evolving separately for tens of millions of years, and then we bring them together. You go to Africa, someone from Africa comes here, you go to Asia, vice versa. We're moving goods around; we're moving people around; we're moving viruses around; we're moving fungi around.

Just to give one really local example of what happens, people have probably heard of a white noise syndrome killing bats all across the United States. That was a fungal disease, a fungus brought over from Europe, probably totally unwittingly by someone, that's just killed millions and millions of bats.

Oftentimes when people try to date when the Anthropocene began, one date they use is the beginning of the use of fossil fuels, and another is postwar, when the human population really, really took off.

All of these questions of what we're going to do to enhance our own lives, unfortunately, are coming at the expense of all these other species with whom we share the planet. And, that is, I unfortunately have come to believe, a pretty inevitable conflict. Because there's only a certain amount of space, and there's only a certain amount of resources on planet Earth.

Just to bring things full circle, and then I'll hand it back over to Marcelo, there are going be the questions, "Well, we killed off this. We killed off that. Can we bring it back? Can we use its genome?" We now are so sophisticated. People have the genome of the Neanderthal. Can we recreate a Neanderthal, just reprogram a human cell to be a Neanderthal, or an elephant cell to be a mammoth? Those questions, if we're all around in a technologically advanced society like today's, a hundred years from now, which I'm not 100 percent sure of, are going to be very, very urgent. And we'll be debating them very, very heatedly.

S. MUKHERJEE: So, maybe this GoPro clone thing is not so bad? Just in one.

E KOLBERT: One. Yourself.

S. MUKHERJEE: Yes, one of yourself.

E KOLBERT: I see. No kids, OK, shrinking population. Yeah.

S. MUKHERJEE: No kids. Yeah, the shrinking population.

E KOLBERT: We could combine forces.

S. MUKHERJEE: Exactly. Exactly.

M GLEISER: I think that's great, because you nailed it. There is a clear tension between these two narratives here. On the one hand, there is science as the conqueror of all maladies, including death, viewed perhaps as a disease to be cured in some way.

Even though I'm a scientist, I have a very visceral aversion to scientism, the idea that science actually has answers to all questions, because it does not, for sure. It's highly debatable, I think, that death is something that we can really deal with, for many reasons. Even invoking physics and the cycle of thermodynamics. In the sense that every time you solve a problem, every time you kill one of the killers, other killers show up. You are one of the best historians of medicine around, and you know this is true. The question is, can this ever end? Is this just part of the way life manifests itself?

On the other hand, as you said, there is the other narrative, which is destruction. We as a species, in order to be so many, and in order to live well, the first thing we need is energy, is environment. And that obviously has a tremendous cost. This is a question I have for both of you, because I have no idea how to answer it: when you talk to people about these things, Elizabeth, when you talk about the sixth extinction and you put the data in front of people, you discuss this with educated people, they understand it, but they don't change their behaviors. Why not? To me, that's mind-boggling.

E KOLBERT: This is maybe not really an answer, but people have been at this project for a really long time, this overcoming of evolution. I think it would be called debated, but most scientists—once again I'm going to defer to scientists—would say humans did in the megafauna. The megafauna extinction was really early on. Humans arrived in Australia fifty or even sixty thousand years ago. There was a wave of extinction of large animals. They arrived in North America probably fifteen to twenty thousand years ago. There was a big wave of extinctions. We no longer have mammoths or mastodons running around New York, although we have plenty of skeletons of them.

It's a project, I think, unfortunately we've been at a long time. It's just that our tools have gotten a lot better. When you think about the pace of change right now, for example your iPhone, everything has been collected from somewhere on the planet and has had an impact somewhere on the planet. The changes that are required when you have 7.6 billion people on the planet—even if you just said everyone needs to eat, that's a basic requirement, these sort of basic human rights, these things come into real serious conflict pretty quickly.

Now, if you said, "Let's do our best, our absolute best. We would like everyone on the earth to have a decent life, and we would still like to preserve as much of the rest of the planet as possible," the best proposal I've seen for that comes from E.O. Wilson. He has a book called *Half-Earth*, that says we should put half of the land surface of the earth aside for other creatures. I think that's a really important idea. But it turns out a multi-national project like that, it's really, really hard to even get it talked about.

M GLEISER: I think it's something to do with the amount of pressure that people respond to at an individual level. Unless there's enough pressure, you won't have gotten to that threshold yet, and you will not do anything. You won't take shorter showers. You won't eat less meat. Whatever it is, even if you know that there is a cost to all that. It's an interesting reflection to understand what it takes for people to do something, not at the governmental level, but at the individual level.

Because we have power as customers. I think there is a whole emerging idea of corporate ethics now, where some corporations are beginning to understand that if you don't like a company and you stop buying their products, they'll have to change. And so you have some companies that you wouldn't expect getting organic products out there and stuff like that.

That's just a reflection that I think we should have. And what about the discussion about the ethics of gene manipulation. Where are we with that conversation?

S MUKHERJEE: Let's talk about the technology first, and then let's talk a little bit about the ethics of genetic manipulation. Again, the background is this reading and writing paradigm. Remember we are now becoming progressively, particularly with computational powers, able to read, to make sense of this encyclopedia, 66 volumes of the encyclopedia. Then comes the question of writing. For a long time, as many of you know, the writing piece of it, changing the human genome perspective, which would be essentially saying, "Go to page 66, and erase the word . . . ."

M GLEISER: A-C-T-T?

S MUKHERJEE: "A-C-T-T, and change it into C-C-T-T." And that might mean changing a gene like a mutation in the breast cancer–causing BRCA1 gene, and making it the non-breast-cancer-causing, so-called normal variant, or the Y type variant.

That project was very badly stuck, for very good reasons. The reasons are very deep, and biologically very deep. They are biologically deep because, as you can imagine, any time you take information away from any genome is a cellular crisis. Cells are extraordinarily sensitive to their genomes being tampered with, for good reason. Because when that happens the consequences are usually quite severe. Cancer is one of the major consequences—a mutation that eventually causes cancer.

We've discovered this in our lab, and many, many labs have been working on this. Cells are extraordinarily sensitive to these kinds of changes, and in general, they were very hard to deliver.

The wormhole—and again this goes back to taking rapid control of evolution—the wormhole came from a very unusual place, and it was not expected. The wormhole came from bacteria, the microbial world. It came because of evolution. It came because in the microbial world, the resources are often so restrictive, there's a constant cat-and-mouse game between microbes to compete with each other.

In the microbial world, what happened many, many, many years ago is that some bacteria began to be invaded by viruses (We've reconstructed all of this, by the way. We don't know if this is true. It's mostly reconstructed). Bacteria began to be invaded by viruses. You can think of viruses as parasites. They use the bacterial cell to make more copies of themselves. They're not living in the traditional sense. They parasitize the bacteria's body, this genetic apparatus, etc., to make more copies of itself. That's what a virus is.

We knew all this, but what we began to understand is that bacteria have a counter-evolution mechanism. They evolved a mechanism, basically, to detect the DNA of the virus and make a cut in it, chop it up.

The field sat there for a long time, this idea that scientists began to recognize, that some bacteria could have a recognition sequence. They could recognize a virus sequence and deliver a directed chop, not to any other part of their own genome, that was protected. It was specific for a sequence.

I'm going to rephrase exactly what I said. In other words, the bacteria could recognize A-C-T-C-C-G-T-T-T and chop that section, A-C-T-G-C-T-T-T. Now, obviously, if you think a little bit ahead, you could say to yourself, "Well, if the bacteria can do that, what if I re-engineered the bacteria, put it into a human cell, and now said, 'Instead of going and chopping the virus, why don't you go chop the human genome, A-C-T?'" As far as the bacterial apparatus is concerned, it can't tell whether it's a virus or a human genome. It has no fundamental mechanism to tell.

In fact, that's what has happened. In the last five years, we have co-opted this bacterial immune system, which is an anti-viral system, to chop up, or to be directed, "Go to page 66 in the *Encyclopedia Britannica of Humans* and go and make a directed change in the human genetic material."

Human cells haven't given up. They are still protesting. They are saying, "No, no, no. Wait a second. I'm not ready." And sometimes they shut themselves down. But the technology is extraordinarily simple. I could tell you that it would take a post-doc in my laboratory on the order of weeks to months to make a directed genetic change in a human stem cell. It now takes two days and virtually all of them work.

M GLEISER: And this is CRISPR?

S MUKHERJEE: This is CRISPR, yes.

E KOLBERT: I have a question. Once I had an infant or an adult, I can't change all of their cells, right? So, I have to go at the moment of conception in there?

S MUKHERJEE: This is a very important question. There are three kinds of changes, and we should know about all three of them. One of them is making changes in all other organisms. This is using CRISPR to make modified crops, modified animals, etc., etc. There you would change, presumably, the sperm and eggs. Actually, you wouldn't change the sperm and eggs, you would change the stem cells that make sperm and eggs.

The big ethical debate there is the debate about diversity, bio safety. That's the central debate. And this, of course, goes very much into things that you think a lot about, Elizabeth.

The second one, the second broad arena, is making changes in human cells, but not in sperm and eggs. As long as they're non-sperm-and-eggs, like blood-forming stem cells, like stem cells that make [the] pancreas, stem cells that make skin, etc. I'm using the word stem cells because you make the change, and you let the cells propagate the change in the body.

There the big safety concern is about possible consequences in the person [who] receives the cells. The biggest one of them being cancer. Cancer, cancer, cancer, cancer, cancer, basically.

Number one and number two are already being done. We will find the first CRISPR based therapies in the next five years. We are working on them ourselves, actively. Changing human stem cells using

CRISPR-like technologies is here. It is now yesterday's new as far as science is concerned.

The human experiment needs to be done. We don't think that they will cause cancer in the sense that people are worried about. Maybe there'll be some. This is a very active area. Lots of debate around this.

The third one, which is obviously the most challenging one, the one that's most debated, is making changes in human sperm and eggs, or rather stem cells that make sperm and eggs. And, if you take that, the third arena, this is different from the other two because this change is permanently heritable through the human genomes. This is a prospective change. Technologically, there's no fundamental reason it cannot be done.

The most surprising piece that's emerged in the last three years, at least to me, is that big groups were put together—the National Academy and National Institute—and we had thought that they would say, "Absolutely not," for human genetic change. In fact, what they said is for diseases that have extraordinary suffering in humans, we should be allowed to explore this arena for human beings. That was the first recommendation.

This will change. This is an evolving conversation, but that was a big line in the sand that we are going to be able to cross. We're going to say that for diseases that cause extraordinary human suffering, potentially a gene like APOE, we should be able to make a prospective genetic change in human sperm and eggs. This is a wide debate. It's not going to close tomorrow, but that's where we're sitting right now.

M GLEISER: It's interesting that this is, I think, the two hundredth anniversary of the publication of *Frankenstein*, which was a cautionary tale precisely about how far science can go without messing things up big time. Everybody watches the movie, but if you read the novel, the novel is about the monster's brain not being Mr. AB Normal's, but actually being a genius, having the brain of the genius, and begging the doctor, Victor Frankenstein, for a companion, a female companion. When the doctor realizes that his experiment first, worked, and second, could create a race of "monsters" [who] are much more powerful, stronger, and brighter than humans, he panics. Hence comes the drama.

The reason why I bring this up is because, imagine that we get there, that we can do this. You can put as much legislation as you want into scientific practice, but it's a bit of a Pandora's box. Once a scientific idea is out in the world, you just can't put it back in. There will be groups that will be doing it for the wrong reasons. Some groups will be doing it because they can pay to do it. And because the therapy maybe very expensive at first, you start having a sort of unevenness in society of those who can take advantage of it and have their kids be super smart, super beautiful, super etc., and those who can't. You go into the *Brave New World* scenario.

S MUKHERKJEE: But, just to temper some of that . . .

M GLEISER: Yes, please. Temper it.

S MUKHERKJEE: As I said, there is a substantial camp of people who correctly believe that tampering with human and animal genomes is going to be fundamentally restricted by the fact that typical traits that we want to influence are run through tens of thousands of loci in the human genome. Basically, you cannot use CRISPR in that sense. You get into a numerical problem. You can't tell CRISPR to change tens of thousands of genes. CRISPR evolved, as I said, to make very directed changes in very directed loci in the human genome.

Now, there is a smaller other camp that says that's not true. It says that, sure, there could be tens of thousands of variations in the human genome that influence a single trait. But that either you could create a synthetic gene, completely outside known biology, or that there's a natural gene variant which will override all of this.

We've seen examples of that. I'll give you one great example of that. Height is controlled by tens of thousands of loci, maybe hundreds of thousands of loci, across the human genome. Yet, one gene, like the Marfan syndrome gene, can override all or many of these and cause extreme changes in human height. I can give you several other examples of this.

This second camp believes that there'll be more examples of that. In fact, perhaps the most challenging example is that in animal studies, we now know that maze-solving cognitive abilities of animals is controlled by tens of thousands of genes. And yet, by influencing two or three of them, you can get a maze-solving animal genius.

I would say the latter camp is a minority. The former camp that believes complex human variation is going to be tough to tackle using these mechanisms is a majority. Elizabeth, I'll ask you a question about that. What do you think about bringing back biodiversity using these new metrics? Are you enthusiastic about that?

E KOLBERT: It's like all roads lead to George Church, the Harvard geneticist who has spoken about bringing back a mammoth. Actually, Asian elephants and mammoths are quite closely related, more closely, I believe, than African and Asian elephants are to each other, and that we could just rejigger an Asian elephant and get a mammoth.

I think that there are a bunch of things to be said about that. Let's just assume for a moment that we could overcome the hurdles, which are great. Because as soon as an animal is extinct and dead, recreating its genome becomes extremely difficult. But let's say we have an animal; we are driving it to the edge of extinction. We take a live animal, we sequence its DNA—pretty easy to do right now—we bank it, and then we decide, "Let's bring this animal back." There are a lot of questions you have to ask. One is, why did it go extinct in the first place? If it's been done in by disease, if it's been done in by an invasive predator, for example, you put them back out there, you're just going to have the same problem all over again. So, you've got to deal with that.

But I think a more profound question, and this actually gets back to your clone with a GoPro, is what is identity? There's a case right now—and this is not sci-fi at all—it's a very simple case of conservation biology. They had a species of the Hawaiian crow that got down to really, really low numbers. It's a distinctive species from our crow here in the continental United States, considered sacred by native Hawaiians, and was down to around twenty birds. They did all sorts of things. You set up a breeding facility. You double-clutch birds. You take the eggs. The birds will lay another egg. People raise the birds. You can do all these things. They brought it back to a point of having maybe 150 birds and just recently put some of them back out in this forest, which they very carefully tried to rid of predators. And the first ones were killed by hawks because they didn't know how to defend themselves from hawks. They hadn't been trained.

An animal is not just a collection of genes. The more we learn, there's a lot of culture—

S MUKHERJEE: The crow needed a GoPro.

E KOLBERT: There you have it!

S MUKHERJEE: A parent crow needs a GoPro to transmit to its [offspring's] GoPro.

E KOLBERT: Exactly. But I'm thinking about my clone. It doesn't want to watch my memories. It's not interested in my memories.

S MUKHERJEE: You would force it.

E KOLBERT: I actually have clones. I have identical twins. So I have clones. And I always wanted them to write about, "I am a clone," but they never wanted to do it. But they're actually very different people. They've had very similar experiences, and they have identical genetics. Not epigenetics, perhaps, but identical underlying genetics. And they're still very different people. So I'm not sure I see the point of all that.

M GLEISER: To a very large extent, we are defined by our memories. The clone may look like you, but it's just a thing. It's not you anymore. Of course, if you go talk to real transhumanists, they're going to say, no problem, in the future we're going to be able to store the essence of your consciousness in a digital program. As long as you back it up, you're going to keep having it. You can always re-implant it in the brain of your clone.

S MUKHERJEE: This is what I was saying. This is the soft version of—

M GLEISER: Exactly—this is GoPro Plus. And now for some questions.

QUESTION ONE: How was in vitro fertilization perceived at its advent? Do you anticipate that using CRISPR to modify embryos will eventually become as acceptable to society as IVF is now?

S MUKHERJEE: That's an important question. When in vitro fertilization was first conceived, it was conceived as a mechanism to get over a terrible burden, which was infertility and people's desire to have children. That was the original driver for that. What's interesting about it, as you very well know, is that it has remained very much in that arena, but it also becomes enabling technology. I'm trying to make a general principle out of this, which is that although it was invented for the purposes of solving an illness, it has become enabling technology for prospective genetic change that may not have to do with illness.

By that I mean that because the egg is outside the human body, you can do things that you couldn't do in the human body, including things that I just talked about before. You could do selections for different traits. And that, interestingly, has nothing to do with CRISPR. That only has to do with reading the genomes. If you can read the individual genomes of in vitro fertilized embryos, you can select one versus the other.

In fact, many of the people who work or think about genetic change say that, as far as constraining the field of human genetic or genomic changes, that's enough. That's sufficient. Unless you want to make changes that are introducing to the human genome things that are entirely foreign to the human genome, you don't need to do very much. By embryonic selection, you can achieve virtually anything that the CRISPR mandate of extraordinary suffering has to do. You could remove any forms of extraordinary suffering, again, given constraints, just by that technology alone.

QUESTION TWO: How optimistic or pessimistic are you with respect to the future of humanity in a planet that is changing so fast?

E KOLBERT: One of the complexities here is that we are so adaptable and so clever and such great tool manufacturers that we are thinking of mucking around with the genetics of everything on Earth, just to name one of an infinite number of examples. So, optimistic, pessimistic? I think it really gets into, are we going to do ourselves in? Quite possibly. We've had the capacity to do that ever since Robert Oppenheimer said, "Now I become death, the destroyer of worlds." We're going on more than seventy years, when we did the first Trinity tests, and that's not a very long time in the history of the planet to have the capacity to destroy the planet. So, the fact that we haven't done it yet doesn't make me that wildly optimistic.

S MUKHERJEE: But . . . There's a moral syndrome that's building around not tampering with human genomes. It's not clear to me that there should be a moral syndrome around not tampering with genomes. What's the morality? Why should we not?

Remember this CRISPR argument came before a panel, and the most vociferous people who argued at the panel were patient advocates, people whose children are affected with terrifying diseases, etc. Just to sound a note, that yes, it sounds horrible—we've destroyed the environment, etc., etc., etc., but on the other hand, I think it's fair to say that scientists aren't doing this because we are fundamentally Frankensteinian, you know?

E KOLBERT: No, I did not want to accuse you personally. And I didn't want to accuse scientists in general. Look, we are all eating GMOs every day, whether we want to or not, and it's feeding a lot of the world. I'm simply saying that we do have so many capacities to change the world on such a fundamental level, from the macro level of blowing ourselves to smithereens and changing the atmosphere and changing the chemistry of the oceans to the micro level, that I sort of feel like all bets are off. I don't know where this is going.

S MUKHERJEE: It's not a GMO vs. go and shop in Whole Foods debate.

E KOLBERT: Right.

S MUKHERJEE: The debate for GMOs across the world is, are there GMO versus CMO, chemically modified organisms, which we've been using, spraying insecticides, which dumps twice as much toxins in the oceans, etc. That's the reality. So the question—

E KOLBERT: I'm going to push back on that. Because a lot of GMOs are modified so we can put more chemicals on them. So, that's a whole other thing.

S MUKHERJEE: Fair enough. So, here's my question to you: do you think it's a complete pipe dream? We have a lot of hungry mouths to feed. That we could feed half of India without the high yielding of rice or wheat is a question mark. What's the other option? What's the middle road? Is the middle road that we make seed banks? Is there no middle road? Are we doomed down this road? Should we be CRISPRing things so that we restrain ourselves? I mean, that's also a possibility, right? We could use CRISPR [to] police. We could be restraining ourselves in a variety of ways.

Is there a middle road, or once we're driving down this road it ends with monoculture?

E KOLBERT: All these things are presupposing a kind of orderliness to the world. Is there a middle ground? Could we use genetic modification to come up with some really great crops, intensely farm certain areas of the world, leave other areas of the world unfarmed and get through to a moment when population starts to fall, etc.? In theory, yes.

Now, look out on the chaos of the world today. We could go down the list and say, "OK, is that going to happen?" I bring to this, I think, an ordinary American's level of skepticism. We have to decide to do things. We cannot decide as a world that we should reduce carbon emissions. Nothing could be simpler. That is the simplest step. We have to reduce carbonization. It's step one. We have technologies—they're simple ones, they're called windmills and solar cells—to start to make a dent in this problem, and we are not doing it.

The idea that we're going to come up with some great construct, yes, it's theoretically possible. Do I think it's going to happen? I'm extremely skeptical.

QUESTION THREE: What about a fourth scenario? We're already immortal. Perhaps life is truly eternal, but in a different form.

M GLEISER: I'm not sure I understand the question very well. Perhaps you mean that life just reinvents itself as time goes by, like the theory of evolution? I don't know.

S MUKHERJEE: I have no clue. Maybe it was a spiritual question.

M GLEISER: OK, so here is a comment about this whole transhumanist idea. If you really take it to the extreme, the real extreme transhumanism is to get rid of the flesh altogether, is to become information, information that can basically just keep going from machine to machine, or eventually not even in any machine. Isn't this just reinventing the soul in modern technological terms? You just get rid of the flesh. The flesh is a problem. Look at all this mess we have to do—cut genes here and all that stuff—and if you get rid of it, all you have is information. If that's really true, then here we are. We just transport ourselves immaterially across time.

There is a huge religious component at the core of this transhumanist movement. Lots of scientists don't, obviously, think like that. But I think it's there for a lot of people [who] actually believe in it and want to feel it.

QUESTION 4: As authors of books, you are immortal. Your books, your words, now live forever. How do you feel about it?

S MUKHERJEE: I mean, I wish I had not put the word *with* on page 66 of . . . I'm just making it up. I don't know. Elizabeth, what do you feel?

M GLEISER: Let's put it another way. Why do you feel compelled to write books?

E KOLBERT: Wow. I'm not sure I do feel compelled to write books. It's kind of a torment to write books. I would argue, for myself, that there's a certain satisfaction when you're done, after the torment. And you are hoping to put something out into the world in a package that you hope will make people think differently. I guess that's what I hope for, and I'm sure, also to a certain extent, what Sid hopes for. As for immortality, I live in a college town and I look around at the infinite number of books that have been sitting there and not taken out for generations. So I'm not sure that that's really a good way to achieve immortality. How's that?

S MUKHERJEE: In my case, I have a very simple formula, or thought, about why to write anything. I think everything should raise a question or ask a question. My books are very much directed by individual questions. The first book, *The Emperor of All Maladies*, came about because a patient asked me a question. I was giving her chemotherapy and she said, "Where are we going with all of this?" And you can take that question at face value, or you can ask the question, "Where *are* we going?" This is an illness that has redefined our relationship to disease. So much so that we no longer use society as a lens to view cancer, we use cancer as a lens to view society. That much influence on our perception of ourselves, our perception of our future, and our perception of our mortality, and yet we had no history, no roadmap, for that.

In *The Gene*, I raise a question about, "Who are we, and what are the limits of what we change?" I think in general, that's been my approach. Elizabeth's book, which I devoured by the way, *The Sixth Extinction*, is a book that raises a question and tries to offer an answer to a question. Those, to me, are fun.

And even fiction does that. My wife's a sculptor and I'll go to the studio, and she'll say, "Tell me what you think about this work." And the first thing I'll always say is, "What question is this piece trying to answer? What question is this painting trying to answer?" Every painting, everything in my lens—it may not be your lens, but in my lens—is about, "What question is this trying to solve? What is it trying to solve? What is it trying to raise?"

It's one way of viewing the world. It's a way of viewing the world in which everything is driven by a series of questions.

M GLEISER: At the end of the day, I guess we're meaning-seeking creatures. We do the things we do because we have the urge to engage with these, right? And that's what makes you, if not completely whole, at least more whole.

QUESTION FIVE: On soft immortality—in some sense, we're already living it. Fifty billion cells from your body die every year.

M GLEISER: Is that correct? Fifty billion?

S MUKHERJEE: It's fifty-two . . . I don't know!

QUESTIONER FIVE: Our bodies days ago are different from today.

M GLEISER: That's an interesting question. Where is the blueprint?

S MUKHERJEE: This is the Delphic boat question. The famous conundrum which has been raised many times by philosophers for a long time. If you have a boat and you replace every plank in the boat, is the boat still [the same] boat? This actually is a question that has very much to do with things that you were thinking about too, the question you raised about identity. If you recreate the planet and repopulate it with everything that we ever saw—

E KOLBERT: Would it be the same planet?

S MUKHERJEE: Would it be the same planet? It is a very fundamental philosophical question. I'm not sure I have a single kind of answer to that, but the fundamental difference is the difference in autonomy. Again, this goes into some things that we've been thinking about as biologists for a while. Why is a body a body? It's a Delphic boat question. If your cells are being replaced all the time, why is the human body a single body? It's a Delphic boat. Where the question begins to become interesting is your body remains a body because it has autonomy. It is servicing yourself. It is in service of something.

When you necessarily create a clone—and again, there are many parallels to the planet here—when you make a clone, that clone assumes autonomy of itself. If there was a war between you and your clone, the clone would side for itself. It would say, "Go take your GoPro memories, and I'll tell you where to put them." This may indeed be the case if we wanted to do this version of immortality, or this version. Even if we take transhumanism to the *n*th extent, as information movement, what is there to say that at the third generation this machine will [not] say, "I'm not interested in your memories. Go take your memories. I want my new ones." We would end up being right back to, ironically, where we are now, when our children say, "I'll tell you what to do with your recommendation to read *War and Peace*, Daddy."

Obviously, these are deeper questions, but they go back, the metaphors extend across human beings into planets and ecosystems.

M GLEISER: There is just one last thing that I wanted to say. You can be sure that, given modern astronomy, there's only one planet Earth in the whole universe. There may be other planets that could be somewhat similar, but no planet will have the same history as our planet. This is the only Earth that exists. Furthermore, there are no other humans in the whole universe. There may be humanoids out there. Who knows? We have no clue. But there's only one species like our own, because we are the product of a very specific, contingent history of our planet. We're very special at the end of the day, us and the planet. So, let's do something to protect it.

8

ON BEING HUMAN

A Dialogue on Literary and Scientific Perspectives

MGLEISER: Being human is a multidimensional experience. We are animals in one sense—we have to sleep, we have to eat, our nails grow, our hairs grow. We have a dimension to our existence that's completely animalistic. On the other hand, we are creatures made of stardust. And we have the privilege to know it. We are made of atoms of carbon and nitrogen, and these atoms are older than the sun and the earth. They came from stars that exploded around five billion years ago or even earlier. We are very, very old stuff indeed, when we talk about what we're made of, the stuff in our bodies. We and all other living beings in this planet. On the other hand, we are also creatures capable of thinking about infinity, of contemplating the divine, of being spiritual, of being creative. We have an urge to make sense of things, to matter in the world and in our lives. This multidimensional experience of being human is the focusing theme of this conversation.

Tasneem Zehra Husain came to writing via theoretical physics. She received her PhD from Stockholm University in Sweden and pursued post-doctoral research at Harvard before returning to her native Pakistan as a founding faculty member at an elite school of science and engineering. She's particularly passionate about the need for a more nuanced, more human paradigm in science writing, one that is truer to both the process and the spirit of the endeavor. Tasneem has conducted several writing workshops for scientists, including an ongoing series at CERN. Her writing has appeared in *Nautilus*, as well as various anthologies of science writing for both adults and children. She's a regular columnist for 3quarksdaily.com and the author of the popular science novel *Only the Longest Threads*. For over a decade Tasneem has been actively involved in outreach. She has worked with K–12 teachers, high school students, and government officials, both in the United States and abroad, and is a frequent speaker at book festivals, science festivals, writing conferences, and physics conferences. She sees humans as storytelling creatures.

T. HUSAIN: I don't know how this evening will go. I don't know what David and Jerry are going to say, but I do know that we will have a fascinating conversation between people who have thought deeply about things from their own points of view but are open to other perspectives. This openness is vital because the big questions, like Marcelo said, are not questions that we're going to be able to answer from any single point of view. We need to bring many perspectives together.

When Marcelo told me the topic for today's dialogue, my immediate response was, "Could it be anything less expansive?" With a subject like "On Being Human," where do you begin? I started writing down all the things that came to mind. The very first thought that popped into my head was that we are storytellers. Since time immemorial, in every society all over the world, stories have been a way to convey our culture and traditions, a vehicle to pass down wisdom through the ages. The Harvard biologist and Pulitzer Prize winner E.O. Wilson said, "The stories we tell ourselves are our survival manuals," and that strikes me as so true.

Stories preserve information by making it emotionally relevant. You remember a story in a way you could not remember a fact, so if someone wants you to be able to recall something in the years to come, it helps if they make up a little story about it. Our ancestors knew that intuitively. Ancient stories are still remembered today. Even simple stories, along the lines of, "Don't go too close to this mountain if it's rumbling because it will rain fire," will be told and retold and are a far more effective deterrent against volcanoes than a warning would have been.

Such stories exist in almost all cultures and are one of the roots of mythology. When we look at the natural world, there are things we understand and things we don't, and we want to find ways of connecting them to make a consistent picture. We ask why things are so, and then create stories to answer our questions—to make some sense of what would otherwise be random. By saying thunder happens because Thor crashes down his hammer, we restore some narrative coherence to an unpredictable life. Of course, there are several variations of the story, since every culture came up with explanations that make sense in their local context. But while many mythic themes are indigenous to a place and its people, it is striking how many are universal.

Of course, myths aren't the only kind of story we have. Ever since people started traveling, we've had travelers tell tales about their adventures in foreign lands. That's always been a way to find out about other places and—especially in centuries past when travel was much more difficult, much slower, and quite rare—designated travelers were sent off on expeditions to find out about other cultures. If you read travelogues from the fifteenth, sixteenth, seventeenth centuries, the authors talk about other people and places as if there's a kind of magical element to them. It doesn't necessarily make sense. Some features are obviously exaggerated and there's often some element of fantasy thrown in.

I grew up in Pakistan, and one of the great oral traditions there is the tale of Amir Hamza. This is an epic that was well-known at the time of the Mughal Dynasty, so it has been around since at least 1500 A.D. It's based on an actual historical figure, Amir Hamza, and his fantastic adventures, some of which ring true and some that do not. When it comes to Amir Hamza's travels, many of the places mentioned are recognizable and there are all these details that make perfect sense—but then suddenly he comes up against a Djinn, or some kind of weird mythic monster. Fact and fantasy are so intertwined, it's hard to separate them. I think this is an extreme case of what happened with travelers' tales. Foreign lands were so completely unknown, and the things visitors saw were so removed from their everyday lives that they often did not have a parallel in the familiar. In the absence of an appropriate vocabulary, travelers projected their bewilderment and fascination into their descriptions, and their accounts reflected both their physical and emotional experiences. Facts they witnessed abroad appeared as strange as fantasy and the unknowns were conflated.

So, broadly speaking, there are stories we tell to make sense of the natural world, and stories we tell about things beyond our experience, that enlarge our vision of the world. That is the first thing that came to mind when I thought about what makes us human: Whether dealing with the visible or the unseen, the immediate or the unseen, we make sense of life by telling stories.

Another quintessentially human characteristic is the instinctive search for pattern. Faced with a seemingly random amalgamation of objects, we have the urge to group things together, to make sense of it somehow, to link individual objects so that there's some kind of scheme to the collection. We organize. We classify. We look for patterns, repeated motifs, any hint of an underlying structure. Even in apparent randomness, we're always looking, subconsciously, for the more predictable, more stable structure we feel must be hiding somewhere.

Just such an exercise led to the periodic table of the elements. A proliferation of elements had been discovered, but they did not necessarily seem to have much to do with each other. Some were obviously related, others appeared completely distinct. There was no understanding why this was so, or how many more elements could be expected. So scientists tried to group the elements they knew in a way that made sense—by atomic weight—and soon they began to sense a structure and see a pattern.

The power of classification is not just to organize or collate what we have that already exists in front of us. The reason we seek patterns is because we want to predict what comes next. Whatever you have in front of you, you could probably group in several different ways, but the real fun comes when you manage to group it such a way that most of it fits, but there are some holes. And then those missing pieces turn up, and that's how you know that you came up with a structure that makes sense.

That's what happened with the periodic table and, more recently, with the standard model of particle physics. All the elementary particles that were known could be grouped together in a way that made sense, but there were gaps. When new particles were discovered, with exactly the right properties to fit into those gaps and complete the picture, it was proved that the standard model was the right organizing structure, because it was able to predict the existence of objects as yet unknown, just on the basis of completing the pattern. It's like a jigsaw puzzle. Say you've pieced together almost the whole thing, but there are six or seven pieces missing. You can predict exactly what each missing piece must look like in order to fit in with its neighbors. And when you find those pieces, you think, "I must have solved the puzzle correctly. This must have been the right way to do it."

To a large extent, that is what we do as scientists. We look at the world around us and organize things into a coherent, consistent scheme. On the basis of the pattern that emerges, we make predictions: if such and such happens, you should expect this to occur in the future. And then we go forth to test those predictions.

Another human compulsion is our urge to fill in the blanks. Think of the standard black box, where you can see what goes in and what comes out but have no idea what happens in-between. The correlation between input and output is something animals are quite capable of learning. In fact, that Pavlovian response is how most animals are trained—if you do this, you will be rewarded; if you do that, you will be punished. The human mind, however, is not content with correlation, it searches for causation. And so we instinctively start projecting explanations onto the black box.

If you see that black box going through the input-output cycle enough, you automatically begin to conjecture what occurs inside. Of course, you could populate that black box with any number of explanations, depending on your intent and purpose. You could make up a story for children; "there are little fairies inside the machine, doing this, that and the other, and that's why things happen as they do . . ." The advantage of such an explanation is that any deviation from the norm can be absorbed quite easily, because fairies are famously whimsical characters; and of course the same holds if you invoke the gods. We all know the gods are fickle. So it could be that everything is going according to plan. Thor is doing what he always does, and things proceed as we are used to, but then a freak thing happens that you can't explain so you say, "Oh well you know, Thor had a bad day, he was in a temper." Stories involving gods and fairies have a certain flexibility. You can accommodate departures from expectation by blaming them on the vagaries of fate—and that is why such a story is not science.

Many scientific processes can be likened to black boxes, in that something evolves into (or leads to) something else, and even though the initial and final states can be observed and measured, the process is invisible—it lies within a black box. So you make up a theory, a mechanism to connect the observations. However, unlike the stories we [already] spoke of, this is actually something you can check. The theory is used to make predictions, which are then verified by experiment. If your prediction doesn't work, you don't get to say, "Oh the machine changed its mind." You know that theory you postulated was wrong.

Often, we have models that work perfectly well and reproduce the data at hand, but as our measurements become more accurate, and our perception of facts becomes more subtle, we become aware of discrepancies that were invisible thus far. Only then do we realize that the internal machinery we had imagined, the picture we had projected onto the black box, was not quite the right one.

Newton's theory of gravity is a case in point. It worked brilliantly in all the situations we used it until a century ago, when increasingly precise measurements showed that the theory explained the orbit of Mercury to a good approximation, but not exactly. Numerically, the discrepancy wasn't large, but the logical gap was enough to seed Einstein's general theory of relativity. So it could be just a tiny difference in the predicted output that causes you to rethink your whole theory and come up with an alternate mechanism—one more streamlined, more modern, more of the moment. And if it has the correct prediction (for now), it is what we go with (for now).

In this way, science is like any other profound story that holds deep truths—it comes in many versions, and as our species grows up, we become more aware of nuance. A story told to a five-year-old is usually pared down to the basics, but when retold to a fifteen-year-old, it can accommodate more shades, more inflections, and the added texture often leading to a new or changed understanding of a familiar tale. The sophistication of the audience determines how coarsely grained the description is. If you're measuring things up to a certain degree of accuracy—say you're interested in telling a cat apart from a dog—then there are only certain details that matter. But should you want to explore more subtle classifications—to determine the specific breed, for instance— you need to upgrade your model. So too, in science. As our knowledge grows, we begin to appreciate subtle distinctions and notice previously unsuspected gaps in the narrative. An exploration of these details can shift our entire worldview.

I think about it as focusing a camera lens. You could be looking at something and see a slightly blurry image where you can make out certain attributes but not others. So you keep focusing in until the image gets sharper, and you see more structure. The process of science is an exercise in sharpening our focus and as a new picture comes into view, our theories change to accommodate it. I happen to be in love with general relativity, but I know it's not the final answer because the theory needs to be modified to incorporate quantum effects. Once that happens, how much will it change the character of what we call gravity? We don't know. We can't say what the next layer will look like, but it will superimpose, not obliterate, all the other layers that came before it.

Talking of layers: Mathematical equations and mythological narratives may be vastly different descriptions of a phenomenon, but we need not think of them as mutually exclusive—they are simply distinct layers of explanation. We never get to see the cogs and wheels of nature at work, so the ultimate truth about the physical universe may well be unknowable, and in the absence of a complete understanding, every unique perspective is of value. No single layer contains the literal, absolute, final truth, but when we hold them together in our minds, we're able to develop a more nuanced view of what we seek.

Have you ever done that experiment with your eyes where you hold up your finger and you close one eye, then you [open that eye and] close the other? Do you mind doing that for a second? It's fun to see it happen. Put your finger right in front of your nose, and close one eye. You'll see the finger in one place. Then close the other eye and the finger will seem to shift. The question is, which is the correct position? The finger didn't shift, so is the left eye right or the right eye? The fact is they're both right in their own way, right? (That's too many rights in one sentence!) The point is, the finger appears to be in different places because your eyes are physically separated, and that separation is what allows you to perceive depth.

It's much the same with understanding as with sight. If you are able to hold several perspectives in your mind without having them conflict or overlap, treating each as a distinct layer, you have a richer life experience and a more expansive vision of the world than you would if you insist on sticking to just the one way—any one way—of thinking.

The fact that things have layers of meaning is something that's intuitively familiar from childhood. You can see it in children as young as two or three as they start to engage in pretend play. You can give a child a banana and say, "Here. I'm calling you on the phone." (I don't know if that'll still work ten years from now when they've never seen an analog phone. Maybe it will have to be something flat and tablet-looking, but you get the general picture.) You can put down a box and say, "This is a house." Even a three-year-old would not question, "What do you mean it's a house? It's a box." Is it a house or a box? They can see that it's a house and it's a box. It's both things at the same time and there's no real contradiction. Literally it's a box, and I'm thinking of it as if it were a house. Very young children already have the ability to see something for what it is literally, but also—and simultaneously—as something else. It is an inborn, innate instinct, and one that I think is crucial because it enables us to create representations.

Eventually we learn to graduate from literal representations, where a physical object (like a box) is being treated as something else (say a house), to mental representations. We begin to extract the essential ideas from their physical manifestations, and that's what mathematical symbols are for. Each can represent a physical concept, a variable you see in nature, a process, an object, the temperature of something, and almost magically, you find you can manipulate all these ideas in your mind using just symbols! The capacity for abstract thought is the root of so much human activity, but in particular the mathematical sciences (which is my background, so I feel this quite strongly) and it comes from pretend play, from learning to accommodate interpretations other than the literal.

When I started thinking about what it is to be human, what jumped to mind was that we're storytellers; that we look for patterns and try to fill in gaps; that we make up theories and that we have the capacity for abstract thought. We've talked about all the others, but that last ability, what does that lead to?

In the bleak trenches of World War I, the eminent physicist Karl Schwarzschild sat, calculating artillery trajectories and contemplating Einstein's general theory of relativity. Though surrounded by unprecedented horror, Schwarzschild was able to recede to a realm that existed at the very boundaries of imagination. Schwarzschild took up Einstein's still-new tale and added his own surprising twist to it—the theory, he found, allowed for the existence of a dark star that would swallow light: a black hole. Schwarzschild died in those trenches not long after, and we are only now, a century later, beginning to build instruments to glimpse shadows of what he saw in his mind's eye a century ago.

This same human capacity is what countless prisoners have relied on to liberate themselves from the cells into which they are bodily trapped—Martin Luther King Jr. and Marco Polo among them.

As I looked at the list I had made, of the qualities I considered quintessentially human, I began to try to make meaning out of this seemingly ad hoc collection—because, you know, that's what we do. And here's what I think:

We are pattern-seeking, meaning-making, curious, questioning tellers of tales. We look around us, and ask "Why?" Our answers seed science and story and myth. We take those answers and run with them, creating scenarios that never existed, and ask, "What next?" We take what there is, we multiply it, reflect it back, make it more. We travel down multiple paths in our mind, become multiple people, live out probable and improbable futures, in possible and impossible places. And that, to me, is what it means to be human. It is the ability to transcend the here and now, to rise above the literal and possible, and partake of the potentially infinite experiences that can be created in unbounded space and time.

M GLEISER: Jerry DeSilva is an associate professor of Anthropology at Dartmouth College. He is a paleoanthropologist specializing in the locomotion of the first apes, hominoids, and early human ancestors, hominins. His particular anatomical expertise, the human foot and ankle, has contributed to our understanding of the origins and evolution of upright walking in the human lineage. He has studied wild chimpanzees in west Uganda and early human fossils in museums throughout eastern and south Africa. From 1998 to 2003, Jerry worked as an educator at the Boston Museum of Science and continues to be passionate about science education. Jerry, please take the stage. Thank you.

J DESILVA: What I want to do is take you on a seven-million-year journey back to the origins of our lineage, because we're storytellers. Tasneem is spot-on here. We are trying to figure out where we come from. I have the great fortune of being able to collect the evidence, the physical evidence we have, that helps us reconstruct the path by which we became human. It's really an extraordinary story that we've been able to tell about ourselves in the last hundred years or so. I'll start in 1871.

In 1871, Charles Darwin hypothesized that humans were most closely related to the African great apes. He did this in *The Descent of Man*. This was almost 150 years ago. He said humans are most closely related to the African great apes, but he had almost no data to base this on. There were no fossils from the African continent yet discovered. Molecular genetics and DNA hadn't been discovered either. And yet he was right. He was absolutely correct that humans are indeed most closely related to the African great apes, in particular to chimpanzees and an animal known as the bonobo, which wasn't even known during Darwin's time. It wasn't discovered by Western scientists until 1933.

Since that time, we have spent a lot of our efforts comparing ourselves to our closest living relatives, the chimpanzees. We've noted all sorts of similarities we share with them, and our differences. For instance, we move in a different way. We move on two legs, rather than on all fours—and that's what I'm going to focus on for the next fifteen minutes or so. In addition, our brains are larger, and with those brains we've developed language and abstract thinking. We've developed technology, although some great discoveries by Jane Goodall revealed that chimpanzees also use and modify tools as well. That was a big surprise. We're the hairless ape or the naked ape. We don't have as much body hair or body fur as a chimpanzee. We have more sweat glands than they have, which helps us stay cool when we run. Some things maybe we're not so proud of. We have more subcutaneous body fat and bigger butts than chimpanzees have.

There are all these wonderful similarities and differences between us and them. But one of the mistakes that we commonly make is assuming that we evolved from something like them or even from them directly. Chimpanzees are not time machines. They're not our ancestors. They're our relatives. They live today. They're our cousins. What Darwinian thinking allows us to do is recognize that we share a common ancestor with chimpanzees that lived somewhere in the ballpark of about seven million years ago, presumably on the continent of Africa.

Now, if this is true, if this is the case that humans and chimpanzees are closely related to each other, then there should be fossils that show us the path by which humans have evolved from an ancient ancestor to what you see in the mirror every morning. These are called the missing links. But we have thousands of them. They're actually not missing. We've got lots and lots and lots of fossils that help us reconstruct our evolutionary story. I want to share a couple of observations about fossils in general before I talk about what we've learned from these fossils.

The first is that every single one of these fossils tells a story. Bones tell stories. And every one of these fossils reveals something absolutely fascinating about our ancestry. There was a new scientific analysis that came out just this morning about Neanderthal teeth, isolated Neanderthal teeth, which were used to predict weaning times. When did mom Neanderthals wean their babies? These are just fragmentary isolated teeth that they based that analysis on. All of these fossils give us wonderful information about our past.

The second thing is that we have thousands of fossils. When I tell that to students, they get a little overwhelmed. People think, "Wow, thousands of fossils!" It is impressive what our science has discovered in the last century. But we're talking about time scales in the millions of years. A million is a thousand thousands. So even if I generously say we have seven thousand fossils, then we only have one fossil for every thousand years of existence. If I borrowed Marcelo's jaw and that became the representative fossil for the last thousand years of human existence, we'd be missing a lot. Every time we find new fossils, it forces us to go back and reevaluate our old hypotheses, which is fun. That's how science works. When we get these new discoveries, we're terribly excited about them, because they allow us to go back and reevaluate some of our old ideas.

The third point that I want to make is that what I'm going to present to you is going to appear somewhat linear. However, what we've discovered in the last decade or two is that human evolution was much more complicated, and frankly much more interesting, than the linear story of a chimpanzee slowly turning into a human that we see on T-shirts and coffee cups. It was much more complicated than that. There were all sorts of different human experiments going on over the course of the last seven million years or so. With those ground rules, let's dive into the human fossil record.

What we've learned through discoveries of just the last fifteen years is that the time period between four and seven million years ago, which we knew almost nothing about in the twentieth century, was occupied by an animal known as *Ardipithecus*. They were found in Africa and they were very apelike in many respects. They had long, curved fingers. They had long arms and short legs. They were very comfortable in the trees. They had a grasping big toe. They had chimpanzee-sized brains. But they were different from apes and more similar to you and me in two respects. First, their canine teeth, their fang teeth, were like yours and mine. They were short and dull and blunt. Second, their pelvis and aspects of their foot indicate that they were able to come down out of the trees and walk around on two legs. At the very base of our lineage, at the very start of this amazing human experiment, what differs between us and our closest living relatives are our smile and the way we walk. It really boils down to that.

By four million years ago, *Ardipithecus* had evolved into a different animal on the African landscape we call *Australopithecus*, made famous by a skeleton named Lucy, discovered in 1974 by Don Johanson in

3.2-million-year-old sediments in Ethiopia. Lucy is an icon for our science. She's absolutely marvelous, but she was not alone. We have lots and lots and lots of other Lucys that we've discovered throughout eastern and southern Africa.

Australopithecus had brains that were slightly larger than *Ardipithecus*, about 20 percent larger, and they had mastered bipedalism. They walked on two legs in much the same way that you and I do. Although we find different species of them scattered throughout eastern and south Africa, and in their different microhabitats they evolved slightly different ways of walking. There were different experiments in walking at this time. New discoveries have revealed that the oldest stone tool technology that we have actually goes back this far to the time of *Australopithecus*. Oftentimes, stone tools are thought to be associated with our own genus, the genus *Homo*, but the oldest ones we have now go back to the time of Lucy. *Australopithecus* was the first stone tool maker.

By two million years ago, we begin to find a different kind of animal on the landscape, the genus *Homo*. They evolved larger brains, more humanlike teeth. They had more humanlike body proportions—shorter arms and longer legs. With those longer legs they expanded their ranges outside of Africa by 1.8 million years ago. For the first time in our lineage, we find early hominins in Europe and in Asia. They controlled fire, and fire's a game changer. Once you have fire, you don't have to sleep in the trees anymore. You can come down and sleep on the ground. It's around these fires that I think we started telling stories for the first time.

In Europe, these members of genus *Homo* evolved into Neanderthals. In Asia they evolved into a population that's known mostly from their genetics, a group known as the Denisovans. But back home in Africa, by three hundred thousand years ago, they evolved into us, *Homo sapiens*. This is an African story. Our lineage is African. Our genus is African. Our species is African.

Homo sapiens [are] different from our predecessors in certain respects. We've got smaller faces than they had and more globular skulls, but not bigger ones. We tend to think we have the biggest brains, but Neanderthals had bigger brains than we have. Denisovans had bigger brains, most likely, than we have. Even *Homo sapiens* thirty thousand years ago had bigger brains than you have today. Brains are getting smaller, they're not getting bigger. Think about that for a second.

We are a creative species. We make stuff. We make art, cave paintings. I have in my hand here a replica of a small figurine carved in Austria thirty thousand years ago. It's a great example of what Tasneem was talking about earlier, where this is just a carving out of sandstone, but it's what it represents that makes this so humanlike. This is perhaps a fertility symbol that was made by humans thirty thousand years ago.

What I want to do now is take us back to the very beginning of our lineage. Let's travel back to that starting point seven million years ago. What got us rolling? What got us started on this remarkable human journey? It was the way we move. We move differently [than] other animals. You all walked into this room today, probably not thinking much about how you move. But what I want to spend the rest of my time talking about is just how unusual upright walking, or bipedalism, actually is.

Bipedalism is an incredibly strange way for a mammal to move, and mammals move in all sorts of fabulous ways. There are mammals that fly, mammals that swim, mammals that cling and leap, mammals that swing, mammals that sprint, mammals that knuckle-walk and climb. There are mammals that hop—those are bipeds, but we're talking about one stride at a time bipedalism. And then most mammals do what the cow does—just move around on all fours, like a goat, sheep, horse, cow, dog, or cat. That's the traditional way to move if you're a mammal, but not us. We move on our extended hind limbs and it's weird. It's a strange way for a mammal to move.

To illustrate how strange it is, whenever people see another animal moving on its two legs we laugh. We videotape it. We post it to You-Tube and 2.5 million people watch it. That's what happened when a bear known as Pedals began walking on its two legs through neighborhoods in suburban New Jersey because of an injured forelimb. NBC news covered this. CNN covered this. When another mammal moves on two legs it's news. That's how unusual bipedalism actually is.

And yet, if we're being honest with ourselves as scientists, we don't know why bipedalism evolved. But we can begin to explore the question by looking at some of its benefits. By moving on two legs, how did our ancestors benefit?

There are many scholars who have found that human bipedalism is energetically quite efficient. We use less fuel moving from point A to point B than other animals do. If we all got up now and walked a mile or so, the number of calories we would have burned could be replaced by a handful of raisins. We are so energetically efficient that we don't burn many calories when we walk. Maybe that helped our ancestors survive.

But obviously, the big benefit of walking is freeing up these things— hands. By liberating your limbs from the duties of locomotion, you can now make things with these hands. And sure enough, early on in the ancestry of *Australopithecus* we find evidence for stone tools. Perhaps even prior to that our ancestors were making other things out of materials that did not preserve in the archaeological record.

Some of the other benefits to freeing up the hands have to do with carrying our infants. When a chimpanzee has a baby, she puts it on her back and navigates through the forest as it clings to her, even while she climbs. One of the interesting observations that Jane Goodall and many other researchers have made about chimpanzees is that they do not let any other individuals in the group hold their babies. It is the mom and the baby for up to six months before she shares her infant with anyone else. It's very different in humans.

If a human tried to put a baby on their back, in the absence of a Baby Bjorn, it would slide right off. You have to actively carry your child. As kids became more and more helpless over the course of human evolution, carrying kids became a more and more challenging task. Imagine you're Lucy, you're an *Australopithecus* three million years ago, and you've got your baby in your arms. You look up in a tree and spot a nice piece of fruit. You really want to get that fruit, but climbing the tree with one arm would be dangerous to you and to your baby. What's the obvious thing to do? Hand the kid off to a helper while you go up and get a meal. That way, you're getting enough resources for yourself and for your baby as well, and when you get back to the ground, the baby is returned to you. But that requires cooperation. That requires trust. That must have developed early on in our lineage as we began to cooperatively raise our kids. The whole idea that it takes a village [to raise a child] probably has very deep evolutionary roots back to the beginnings of upright walking.

With our freed hands, we can also carry food. Lots and lots and lots of food. Freeing the arms allows us to do this. But if you have a surplus of food, you're not going to eat it all. You can begin to share that food. Food sharing became an important part of this story as a result of our ability to collect more than we need in our open arms.

Stanley Kubrick thought he knew why the hands were liberated in our ancestors. He crystallized the idea in film, of humans as the killer ape. We had freed our hands to use weaponry to kill members of another group or to kill prey. This has a story to it, and I want to share this, as it is consistent with what Tasneem already spoke about regarding World War I.

During the war there was a medic named Raymond Dart. Raymond Dart was in those trenches and saw the results of the war firsthand. And then he got a job as an anatomist in South Africa. In South Africa, he excavated a fossil site called Makapansgat in the 1940s. What he found at that site were smashed animal bones. He envisioned early human ancestors using these bones as weapons. Using them to stab one another. Using them to attack other animals. We were the killer ape. He published this idea in the early 1950s and an author by the name of Robert Ardrey took hold of the idea and published a book in the 1960s called *African Genesis*. That book became a *New York Times* bestseller. That idea was visualized at the beginning of the film *2001: A Space Odyssey*, in the dawn of man sequence. What made us human was the freeing of the hands to use weapons.

Who could blame these researchers and who could criticize this characterization of human nature, post–World War I, post–World War II? However, in the 1970s, when researchers went back and examined these bones from Makapansgat, they discovered that the bones weren't smashed by humans. They were chewed on by hyenas. Hyenas had broken the bones. It was a misinterpretation of the data that led to this narrative of humans as the killer ape.

I want to turn that idea on its head for the final piece of my presentation and start talking about some of the costs of bipedalism. Because yes, there are benefits to it, but there are also enormous costs.

The fastest human being on the planet is a guy named Usain Bolt. Bolt can run twenty-eight miles an hour. Twenty-eight miles an hour! Whoa, right? Not really. That's only half the speed of an antelope, half the speed of a lion, half the speed of a leopard. He couldn't outrun a leopard. He couldn't catch a rabbit, a squirrel, or a chicken. Humans—and he's the fastest one we've got—are exceptionally slow animals. By evolving bipedal walking and running, we have sacrificed the ability to gallop. We've lost our speed, making us vulnerable out there on the landscape.

Being perched up on two legs rather than being on all fours makes us incredibly unstable. Humans fall all the time. When's the last time you saw a squirrel just trip and fall? Humans are slow, we fall, and we were preyed upon.

At some of those same sites where Raymond Dart was finding what he thought were the remains of kills we had made, researchers have more recently found the remains of early humans who had been killed by hyenas and leopards. We were not the hunters; we were the hunted. Bipedalism makes us vulnerable.

What I want to end with is a story that starts with a fossil femur. It's a left upper leg bone, an upper thigh, that was discovered in Kenya in 1973 by a paleontologist named Bernard Ngeneo. This fossil came from an upright walking human ancestor. We know that because the head of the femur, where the hip joint is, is offset considerably from the attachment for hip muscles, which would make them quite mechanically efficient, give them leverage, so that every time this individual took a step they could balance on a single leg. Only humans have this anatomy.

It belonged to an individual who had already reached full size, because there was no growth plate. At this small size, it's thought to be a female. So, here we have a female, probably around eighty or ninety pounds, a little bigger than Lucy. The volcanic ash around this fossil tells us that this individual lived about two million years ago and it's presumably an *Australopithecus*.

But what I want to draw your attention to is some unusual anatomy. There's a bulge in the shaft of the femur and a thickening of the cortical bone. This is consistent with an individual today who experienced a spiral fracture of the femur. This is a healed fracture. This poor individual, maybe she fell out of a tree, or maybe she was running, and she stepped in a hole and turned that leg. Either way, she shattered her femur two million years ago. There were no hospitals. There were no doctors. There wasn't even controlled fire. There were no homes. She should have died. A leopard should have come and eaten her. But that didn't happen. She continued to survive. Other members of her group must have taken her and brought her up into a tree. It would take six weeks for this bone to heal. And heal it did. The callus shows that she lived for many more years.

Bipedalism gives us all of the things that we recognize as characteristics of being human. The freeing of the hands allows us to be creative. It allows us to make our technology. It allows us to do all the wonderful things that we do, but as bipeds we are also incredibly vulnerable. What I think is important to recognize and end with is that by being a bipedal animal on this landscape so long ago, the only way we could have survived is with this other wonderful thing we have as humans—care, compassion, empathy, and pro-sociality. We take care of each other. We don't leave individuals for dead. That separates us from many of the other animals out there. And I think that has deep origins right to the beginning of our lineage. Right to when we took our first steps. Thank you.

M GLEISER: David Grinspoon is currently at Dartmouth as an ICE [Institute for Cross-Disciplinary Engagement] Fellow. He's an astrobiologist. For those of you who do not know what that means, it means that it's OK now for scientists to think about extraterrestrial life. You actually get grants to do it. It's a wonderful time. He is an award-winning science communicator and a prize-winning author. His newest book is called *Chasing New Horizons: Inside the Epic First Mission to Pluto*, co-authored with Alan Stern. He's a senior scientist at the Planetary Science Institute in Washington and an Adjunct Professor of Astrophysical and Planetary Science at the University of Colorado. He's involved with several interplanetary spacecraft missions for NASA, the European Space Agency, and the Japanese Space Agency. His book *Earth in Human Hands* was named a Best Science Book of 2016 by NPR's *Science Friday*. His previous book, *Lonely Planets: The Natural Philosophy of Alien Life*, won the PEN Center USA Literary Award for Nonfiction.

D GRINSPOON: What Tasneem said is right. We really need the perspective of all of these different viewpoints so we can get that sense of depth and at least begin to address this question of what it means to be human.

It does sound strange, doesn't it, that scientists are getting grants to study extraterrestrial life now. But we do that largely by studying Earth and its history, the way life has co-evolved with our planet, and the limits, variety, and story of life here. We also study other environments in the universe, and then we try to map our ideas of biology onto these other worlds to understand the potential for life elsewhere. That, in turn, reflects back and illuminates a lot about our own planet and our own story. My particular take on this question of what it means to be human is to look at how we, our species, fit into the story of our planet.

I was a child of Apollo. I grew up enthralled by the early space program. The first missions to Venus and Mars and other planets were happening when I was a kid. It captured my imagination, and somehow, I managed to follow that fascination into a career as a planetary scientist and an astrobiologist.

I've been fortunate to be a member of several teams of scientists and engineers who have imagined, proposed, built, and sent spacecrafts to the other planets. We get quite attached to these spacecrafts. In a way they're like our children. We prepare them as best we can and then we send them off on their journeys hoping they'll be OK, hoping they'll stay in touch. And sometimes they don't. But when they do, they send us back information, pictures, stories. "Travelers' Tales" is actually what Carl Sagan called his episode in *Cosmos* about planetary exploration, which I thought of when you talked about travelers' tales, Tasneem.

Through this venturing, this hunting and gathering of planetary stories, we've developed what we call comparative planetology, where we use the similarities and the differences between worlds to construct a general understanding of how planets work, including how our own planet works.

Here's an example of comparative planetology. Consider Venus, Earth, and Mars. I love this trio of worlds, because Venus and Mars are our two nearest neighbors, not just in physical proximity. In many ways, they are also the two most similar planets to Earth. And these three have divergent stories.

If you look at river deltas on Venus, Earth, and Mars, you can see the similarity in form. And yet they're each different in the details. They tell us about the differences in the stories of these three worlds. In the case of Mars, it had rivers early on that dried up. We see ancient flood valleys and ancient river deltas, but they're billions of years old. In the case of Venus, there was a runaway greenhouse, and it became so hot and so dry that these rivers were carved by lava, yet they follow that same form. In the case of Earth, you can see the influence of life in a river delta, which has a complex interplay of the geological and the biological.

These planets, it turns out, started out with very similar conditions when they were all young. As far as we can tell, they each had oceans and less extreme climates (from a terrestrial perspective). Yet they all evolved in radically different directions. Both Venus and Mars experienced climate catastrophes when they were young. Mars froze over. Venus had a runaway greenhouse. Earth went through its own kind of radical change. Earth came to life, and thus it started down a path very different from those of its neighbors. It went through a branching point where life formed and then kind of took over or permeated the entire planet.

I'm going to fast forward the story here, because it's a 4.5-billion-year story, so that's quite a compression ratio. But then, much more recently, something else happened to Earth, something very strange—the appearance of what we sometimes call civilization, although that word is fraught. We find that, in fact, many of the words we use to try to describe ourselves are fraught. But whatever you call it, this was certainly a major transition, and a new kind of transition on this planet.

As an astrobiologist, I'm driven to wonder if this is a transition of a kind that can occur elsewhere in the universe. If you were an alien astrobiologist watching our planet, an alien astrobiologist with a really long attention span, so that you were watching our world in time-lapse over four and a half billion years—and by the way, on that time scale, Jerry, your seven million years is really tiny, a fraction of 1 percent, so that's a short story in this novel—but if you were an alien astrobiologist watching our planet over the eons, you certainly would have seen a lot of changes. You would have seen the tectonic plates shifting around, the continents merging into supercontinents and splitting apart and morphing around in this sort of shifting spherical jigsaw puzzle. You would have seen the climate change and oscillate between a hot greenhouse and ice ages. You'd see the polar caps growing and shrinking, doing their dance sort of quasi-rhythmically over time. Throughout all those changes, all those billions of years, the night side of the planet would have remained a nearly unbroken blackness, with only the occasional flash of lightening or splash of aurora. Then recently, just four hundred million years ago, you would have started to see forest fires at night, as the continents became green, but mostly this side is just an unbroken blackness until very recently, just a blink of an eye in this story. Whoa, what is this? The night side lights up in this strange new way, starting in a few coastal areas and then spreading along these nodal lines across the continents with a pattern that seems sort of organic, but with some other quality as well. Then you would have observed a raft of other changes starting on the planet: changes in the atmosphere; changes in the composition of the ocean; the geometry of the patterns on the land; perturbations to all the chemical cycles—the carbon cycle, the nitrogen cycle; these strange new linear waves crossing the oceans; and then strange linear clouds streaking through the air. Then, very recently, in just the last twitch of time, in the last seventy years, which is really nothing in this fast-forward view, you would have seen something else extremely unusual and unprecedented: little bits of the planet started leaping back off into space. Small insect-like contraptions launching out and buzzing around nearby space and some of them flying off to other worlds and sending radio signals back. Some new kind of technologically enabled curiosity is at work. If you were this alien astrobiologist and had been watching all of this, you would certainly notice that something completely new is happening on this planet, something unprecedented that has never happened before. A force has awakened here. What is it?

You've probably heard this word: Anthropocene. The idea of course is that we've entered a new epoch of Earth history defined by this new force: the combined activities of humanity. If you look at the numbers, the magnitude of changes, it is indisputable that it is now greater in its effects than some of the other geological forces.

There are many ways to describe this, but as a planetary scientist, I'm very interested in the question of what this new change agent really represents in terms of the history of the planet. What's really new and different here? After all, we are not the first species to come along and change the planet radically. We're not even the first species to, in the quest for a new energy source, pollute the atmosphere so much that it causes environmental disaster, mass extinction, and climate catastrophe. That's been done before. In fact, these little guys did it. These cyanobacteria, they look innocent enough, don't they? But two and a half billion years ago, they discovered solar energy. These guys perfected photosynthesis and they exploited it so fully that they polluted the atmosphere with a dangerous gas. That gas was O_2, oxygen, which at that time led to mass extinction.

Now we've evolved to use those powerful reactions of oxygen with organic matter—the same reactions that were deadly when oxygen first appeared. That's how we power ourselves, that's respiration. Evolution is opportunistic and we've made lemonade. Yet at the time, oxygen was a catastrophe. When you hear this story, do you say to yourself, "How could they? Those irresponsible cyanobacteria! They were not good planetary citizens." But we don't say that, because they're just bacteria. And yet today, we see ourselves behaving in a somewhat analogous fashion. Those of us [who] are paying attention and have a soul, say, "Wow that's horrible. That's horribly irresponsible. What are we doing?"

So what's really the difference? What have we got that the cyanobacteria didn't have? Which is really kind of another way of asking this fraught question of human exceptionalism. What makes us different? If we're going to name a geological age after ourselves, is that just an exercise in self-aggrandizement? Is it just arrogant? Or is there really something new going on worthy of a new epoch? Maybe considering the difference between the cyanobacteria's planet-destroying role and our planet-changing ways can help us get at the answer.

There are many ways in which our planet has changed if you look over the whole history and many sources of catastrophe. There are random events like asteroids, which hit the planet, or volcanic eruptions, which caused mass extinctions. Even life itself—as in the case of the cyanobacteria—several times evolution has led to the success of some forms of life that have caused disaster for other forms of life. We are not the first agents of planetary change. Yet I think that word *agent* or *agency* carries a clue as to what's really different here.

In my view, what's really new about the Anthropocene, about this transition that's happening to the earth now, is the role of cognitive processes. Now, somehow cognitive processes have become planetary processes. And abstract thought, our ability to project ourselves beyond our immediate time and place, to form cultures, and then also to use technology to increase our reach and start to change our environment, all this is ultimately an outgrowth of our cognitive skills. Somehow, now the power of that cognitive activity has grown and increased to the point where it's become a planetary process.

But I also think there are two different forms that this planetary-scale activity can take, and I think the distinction between them is very important. I will call them inadvertent planetary changes and intentional planetary changes.

What I mean by inadvertent planetary catastrophe or planetary change can be represented by a picture of traffic. Again, I think that the question of agency is very important, the scale of agency. For example, in traffic, each car is driven by a person with agency. We're very good at that. If you think about it, it's amazing how well traffic works, that we're not always colliding with one another, that we can steer and avoid obstacles and hit the brakes, even unconsciously when we need to. You can see the agency in each driver, yet if you look at the whole system and if you look at the global transportation system, you can ask, "Who's driving that?" And the answer's kind of, "Well, nobody." So traffic is an example of how we participate in systems over which, on a larger scale, we have no sense of agency. The scale of control here is small. It's the size of cars. Yet whole traffic pattern is part of the global transportation system, a planetary-scale activity.

If we look at ourselves on this larger scale, it's almost like that dream where you find yourself doing something [that] you are helpless to change. You want to run from danger, but you can't. What we see here is an example of a species that's really good at solving local survival problems and using technology to extend its ability to do so, but in so doing has inadvertently created a global change that they've sort of stumbled into. That's what I call the Anthropocene dilemma, where our influence exceeds the scale of our agency, or our sense of control.

An example of that inadvertent kind of planetary change, obviously the one that we're all most aware of now, represented by the Keeling curve, is the steady oscillating increase of carbon dioxide over the years. You're very familiar with this and all of the consequent effects, such as the frightening decrease in the polar sea ice, so I don't need to spend time convincing you. This is obviously the most pressing example right now of one of these inadvertent global changes. When we started using fossil fuels and using internal combustion engines, nobody said, "Hey let's change the planet!" We were finding really good solutions to local problems, but the global-scale activity leads to unintended consequences. It was inadvertent.

Another example of one of these inadvertent changes is the ozone hole. You've heard about this as well, which, by the way, was first discovered in part because we were studying the planet Venus. Some scientists noticed a puzzling lack of oxygen in the upper atmosphere of Venus, and they figured out that chlorine destroys ozone. Some other scientists read that paper and said, "Oh that's interesting. What about all this chlorine we're putting up in the stratosphere with these refrigerants, these CFCs [chlorofluorocarbons]?" And they thought, "Uh oh, wait a minute . . .," and put two and two together. And they sounded the alarm.

And the ozone hole gets deeper throughout the 1970s and the 1980s, but then interestingly, it levels off by the year 2000. That leveling off is a really interesting phenomena that I use as an example of the other kind of cognitively driven planetary change, what I will call intentional planetary change.

Because what happened with the ozone was that we saw what we were doing. The response wasn't immediate. There was a big argument. People said it was a hoax. Corporations defended their economic interests and resisted change, a lot of what's going on now with the fossil fuel issue. And yet, the truth became overwhelmingly apparent. Eventually even the DuPont Company got onboard and said, "OK we've got to create replacements and phase these CFCs out." (Of course, before they did that, DuPont patented the replacement chemicals, so it was not all selfless.) Nonetheless, it's an example of where a problem was recognized, there was a global conversation about this, global action was undertaken, and indeed now the ozone layer is on track to being fixed. It takes fifty years for the chemical reactions to bring it back, even if you stop the dangerous behavior, so we've got to stay on task. And there are some wrinkles in that story you may know of, but the overall story is one of success.

This is an important proof of concept. It shows that we have within us the capacity to interact in a different way with global-scale problems. It doesn't mean it's going to be easy. It doesn't mean, "Oh no problem, we got this!" There are many ways in which the ozone problem is a much easier problem to solve. Certainly the economics of it are easier than the fossil fuel problem, which is so deeply embedded in our economy and some massive economic interests. But nonetheless, it serves as an example of a different way of responding to planetary-scale problems.

There are other examples of intentional planetary changes, and I'll give you just a few, but the most important one, of course, is our ongoing effort to change our energy sources so that they do not destroy the natural systems upon which we depend. This is an effort [that] is underway. And obviously it's too little too late to avoid damage. There's going to be damage. Yet it's also true [that] there's now a global conversation happening about this that was not happening twenty years ago, and that there are a lot of signs moving in the right direction—solar energy getting cheaper, a lot of powerful interests getting onboard. I think most of us are painfully aware that this transition is happening in a slow-motion fashion. It's complicated and there are counterreactions and we could spend hours talking about just that. But it is an example of what I call a "global change of the fourth kind," an intentional planetary change, which is now underway.

In my view, we will get through it. One hundred years from now we won't be using fossil fuels. We'll have made the transition far too slowly.

We'll look back and say, "Look at all the damage we did. Look at all the pain and displacement. How could we have been so stupid?" And yet we will move through this to another time.

Looking further into the future, there are other kinds of threats that we can meet with intentional planetary changes. On longer time scales, Earth has been hit in the past by asteroids and comets. It will be hit again. These can cause mass extinctions. Except it doesn't have to be that way, because now we have in place an observation program. We're identifying these objects and we actually know how to stop them, or we think we do.

It may be that a more positive way of looking at the Anthropocene, if we can get over this difficult century we have in front of us, is that this may be the time when Earth is no longer threatened with extinction by some forces that have plagued our planet throughout its history.

On even longer time scales, there's the possibility, the inevitability actually, of dangerous natural climate change. We have this illusion that left to its own devices, the earth is a paradise, and we just have to take our hands off of the scale and everything will be fine. That is an illusion borne of the fact that our current civilizations have grown up in this ten-thousand-year period of relatively warm and stable climate that is actually an aberration in planetary history. If we wait long enough, there will be another ice age. Our current civilization wouldn't survive it, a lot of other species wouldn't either. It's an interesting question: If we know about these longer-term threats and we can conceive of how to mitigate them, then do we have a responsibility to do so?

Even though we're talking about extremely long time scales now, I think that learning to think on geological time is part of the transition into realizing that we are a geological force, of thinking of ourselves as a geological process, of understanding how we're embedded in the history of the earth. And this, in addition to our immediate responsibility of ceasing our vandalism of the climate, I think inevitably leads us to consider these longer-term responsibilities and ponder what role we will ultimately want to play on the planet.

When I think of the long-term future, I can't help but think of SETI, the Search for Extraterrestrial Intelligence [Institute]. People who do

SETI have long been aware of the fact that when you do the math of SETI and you ask the question, "Is there anybody else out there and what are their stories? And what stories might they be able to tell us someday?" the answer is tied to the longevity of civilizations. If all civilizations that discover powerful technology flare out after a few hundred years because they just can't handle that power and responsibility, then there's going to be nobody out there to talk to. You can actually show this mathematically. But what if, on the other hand, some—and it could even be just a small fraction—figure it out? They get a handle on themselves and figure out how to use powerful technology as a tool for survival as opposed to a threat to survival, and they last for tens of thousands, hundreds of thousands, or millions of years, even. Then the math shows that the galaxies should be much more populated with conversant, sentient creatures who we could discover and maybe even communicate with someday.

Then you realize that the exercise of SETI takes on this sort of hopeful aspect: What we're searching for are the survivors. If we hear from someone or discover someone, it means it's possible to get through this little bottleneck we're in now of what has been called technological adolescence. It's possible to achieve great longevity. The truly intelligent species will have great longevity. Are we truly intelligent? We don't know. In fact, our alien astrobiologist I mentioned at the beginning, she's not sure. She's withholding judgment on the human race. She's got a long attention span. She's going to wait and see about us. But it's very clear that if we do want to survive and be one of those long-lived civilizations, we're going to have to change our relationship with the planet and have that intentional conscious mode of interacting with the planet come to dominate over the inadvertent random mode of interacting with the planet.

If we want to create what I call *Terra sapiens*, which is a world in which we've learned to integrate gracefully our activities with natural global cycles, then we have to become a new kind of entity on this planet. We need to learn to live comfortably over the long haul with world-changing technology. We need to use our knowledge of the way the earth works and our newfound awareness of our own influence to gracefully interact with the great cycles of planet Earth. We have to learn to work with the earth, not against it. The key right now, or one key, is the propagation of a world view that is global and multi-generational.

There's one other thing that this exercise in thinking about the Anthropocene as an astrobiologist has made me feel. We need to have a little more sympathy for ourselves. We're trying to do something that has never been attempted before, at least not on Earth. One hears these days a lot of anti-humanism in the responses to the Anthropocene, like, "People suck," and "Can't we just exterminate all the humans and the earth would be fine?" I too am quite alarmed at the direction we've been taking on this planet, but I don't actually want to exterminate all the humans. I cherish art and dance and music and scholarship and all of these things. Yes, we've got some problems, clearly, and some challenges. We've got to grow up in a hurry. When you think about the metaphors we use to talk about ourselves, a lot of times we're a cancer, we're a virus, we're criminals on the earth, we're raping the rainforest. And you know there's a truth to these metaphors, a horrible truth, yet they don't tell the whole story. Because a virus doesn't stop and say, "Hey, maybe we shouldn't be a virus," and have conversations about it. There's something else going on here.

One thing I'm interested in is considering and promoting some other kinds of metaphors. The metaphor of childhood. We're like an infant, first looking at her hands and noticing, "I have agency in this world, I have power, but I don't know what I'm doing here." Or the metaphor of the unschooled driver. You wake up and you're driving down a road and you're driving a big rig, barreling down the road, but you don't know how to drive. And everything and everybody you love is in the back, so you'd better learn how to drive in a hurry. That's kind of us on the earth, realizing that we have this global agency, but we don't have a survival manual. We don't know our story. We need it but we don't have it. Or another metaphor: sleepwalking. You wake up and find yourself in the middle of committing a horrible crime. You don't know how you got there, but now you have to take responsibility and deal with it. I have sympathy for the human race, and I want to suggest that we see ourselves as immature and confused, not constitutionally evil.

"What makes us human? What are we?" From a planetary history perspective, we are the species that can change the world and realize what we are doing. Then the question is, "What are we going to do with that knowledge?" Can we integrate it in a conscious way and change our behavior to become well integrated in a positive, constructive relationship with the planet? This is, I believe, the framing of our current challenge that comes out of thinking of our role in planetary history.

M GLEISER: Three very different perspectives on the single topic of us. We always want to know who we are and what we're here for and what gives us meaning in life. And there are different ways to achieve this.

One thing that occurred to me, trying to bring it all together—Tasneem, you mentioned that children look at a box and they see the box, but they also see the house. They can live with that superposition of images and can have fun doing that. Grownups seem to lose that a little bit. When we look at a forest, we can think, "Let's go forest bathing. Let's go into the forest and replenish ourselves and be one with nature, etc." But we can also think, "Whoo, timber!" Usually, these two groups tend to go against each other. We are storytellers, but sometimes we look at the same thing and we tell very different stories.

I think the way we evolved, if you look at different cave paintings from different times, you know a little bit about the culture of that particular group and how it depicted perhaps the animals or where they were living in different ways. A lot of what we do is contingent on the tribes we belong to or on the groups we belong to. I think the Earth issue—which is really the big issue here because it's not just our issue, it's also an issue for the future generations—is one that, as Dave said, needs a different story. How are we going to do that? What is it going to take to go beyond the tribal walls that we have erected between us, and that make our current life choices really hard to sustain on a fragile planet? Because the planet—or its current life-supporting state anyway—is fragile, and there's only so much abuse it can take from us. What would it take to look at the science, to look at the storytelling, and to think that we need a different kind of myth for our age, a unifying myth that goes beyond tribal divisions and is really a story of our species as a whole? That's a tough question, I know.

D GRINSPOON: Carl Sagan used to talk about identification horizons, how you can go from just thinking about your family to thinking about your village or your tribe and then your country. He used to talk about the goal of expanding our identification horizon to the whole world. And there's this fantasy that's been in a lot of science fiction, or just thought experiments, that what if we did become aware of an extraterrestrial civilization? Not one that was necessarily threatening us, although that could be the story—that is the story in a lot of science fiction—but would that congeal this global identity?

I had a dialogue recently with a religious ethicist from the group at Princeton, the Center for Theological Inquiry, who put a little bit of a damper on my idea that all we need is this global view. He said, "Well, what about ExxonMobil? They have a global view. And they see timber, not the forest." That got me thinking that the identification horizon needs to be not just in space, but in time. Because if you are thinking of future generations and of this long-term presence on the planet, then you're much less likely to think, "Oh, let's just chop down that forest because it's timber," because you're going to be concerned with it. Maybe we need to expand our awareness in four dimensions, not just three.

T HUSAIN: I also think it's interesting how much our horizons have grown. Long ago, when travel was cumbersome and infrequent, people mostly just knew their tribes or their villages, and that belonging shaped their identity. But now, when travel is almost a fact of life, many of us have developed more complex, multi-layered identities. While still identifying with our place of origin, we take on some of the habits and attitudes of each new culture where we live or study or work. And so we develop a more nuanced sense of identity and begin to realize that you can have more than one loyalty. Historically, outsiders were viewed with suspicion and caution—conventional wisdom was "If you're not like me, you are against me." Differences were a source of conflict. I think that's something we're all trying to reconcile in each of our lives as individuals; we're struggling to replace that exclusionary mindset we inherited with a more open, flexible concept of belonging that allows us to honor different traditions and feel rooted in different cultures, simultaneously.

And it doesn't necessarily take all that much, or that long. In fact, I just felt my attitude shift sitting here right now. Jerry completely altered my perspective on something by telling a story that was different [than] what I expected. Toward the end, that part about the healed bone, and what it meant, was so moving. For decades I've heard the old tale about how competitive we are, how that is an essential part of our nature, how crucial it was to survival, but the interpretation Jerry put on it just now changed all that. Suddenly I was able to see us as caring, collaborative creatures who have always been that way. That instinct, too, is part of us. To have that long-term view, that sense of history helps. It's one thing to feel you need to learn an entirely new skill and a completely different thing to feel that you need to remember something you already know.

We've always looked after our own, it's just that our horizons are growing. This whole planet is our village now, our tribe, our family, and we need to treat it that way.

J DESILVA: I think you're spot-on there. It certainly has, as the world becomes more connected, been easier to talk to colleagues around the world to quickly share ideas. I see it in my students. They're more connected with one another around the world than I was as a student. It gives me tremendous hope.

I suppose there is a cautionary tale for me as I look at the past and look at our successes as a lineage. Generally, what we see is that *Homo erectus* is the lineage that really takes off and does well. It travels around the globe and goes to Asia and Europe and Africa and begins to live in different habitats. For a long time, we were a species, or a lineage at least, that was limited to certain places and eating certain things. *Homo erectus* became this ultimate generalist and that's what led to our survival. We would have gone extinct if they had been too specialized.

It is risky for an animal to be too specialized—eventually environmental change pulls the rug out from their specialized adaptations and they go extinct. These beautiful, marvelously adapted animals usually end up going extinct. We see it happening right now with something as magnificent as a polar bear—beautifully adapted for its environment, but look what happens when the environment changes.

Throughout history, apes used to be more plentiful. There are times in our past, twelve to fifteen million years ago, when there were dozens of species of ape that we have in the fossil record, and now there are just a few. They were decreasing in number long before we had anything to do with it. Apes are very specialized. They're too specialized. But one ape, a bipedal ape, became a generalist. As a generalist, we can live anywhere, eat anything, survive anything. A lot of that has to do with our culture and our innovation, but it means we have become incredibly good at extracting resources out of our environment. I look back in the past and say, "Wow look at *Homo erectus* extracting resources out of its environment, what an amazing ancestor of ours!" but then it's, "Uh oh . . ." because that's what we inherited—this tendency to strip resources from an environment. Making decisions to treat the world differently is going against millions of years of our biological tendencies. It's going to take self-reflection and it's going to take a conscious effort on our part, because it's not going to be easy. If this was easy, the environmentalist movement of the 1970s would have changed everything and it hasn't. This is going to be hard. I think it's doable. But it's hard.

M GLEISER: It has to become enacted at the level of individual agency. I'm convinced that it's only going to happen when each person takes responsibility for his and her life and acts accordingly, then talks to the family, talks to the community, talks to the school. It's going to be a grassroots thing. Because if you expect a top-down change of order, it's not going to happen. It's up to us to, kind of the cliche, be the change [we] want to see happening in the world. And it starts with the individual. You have a phase transition—different individuals start to act in certain ways and together make a huge change that really transforms ice into water and other things in physics.

D GRINSPOON: There are examples from our history of those kinds of phase changes. I think what you just said, Jerry, is very sobering. There's definitely some truth to it, but there are also examples of radical reinvention in our long-term evolutionary history, as well as changes in consciousness—behaviors that we all thought were OK, but then the mass consciousness changes, and they're no longer OK. That's the level of change that we need to promote and seek now.

M GLEISER: And now some questions.

QUESTION ONE: I just have a question about your audience. Is ICE speaking to young people and high schools? I mean, you make bones and everything very exciting and by telling stories, you're going to capture kids. The rest of us are on our way out you know. How can we get young people?

M GLEISER: That's an excellent question actually. I have to say that I've been a professor for a very long time, and it's been historically very difficult to attract students to conversations like this. It's not just us, it's pretty much everybody, because they're kind of overwhelmed and when they do have some time off, they don't want to go and sit and listen to another lecture unless it's a superstar person that speaks directly to them. In some places you see much, much younger people. Not all is lost. And David and I do a lot of public understanding of science and there is a way of, say, hitting the young, so to speak, which I think we both do, and which works out quite well. There are different ways. It's usually related to media, social media and stuff like that. For example, I started giving free lectures on my YouTube channel and they're all young, mostly. Those are perhaps different ways of doing it. But you're right, it is hard.

D GRINSPOON: If I can just make a quick comment. When I go around and give talks at museums and places where there [are] more multi-generational audiences, I'm very encouraged by the young people in this country and the way that they're hip to these ideas and these concerns. Obviously, we need to keep educating, keep talking to them, but strangely, right now it's the older people I'm worried about, and the way they're voting. I was talking to friend of mine earlier today who's an environmental activist and we were saying, "God you know if we could just keep the world together and hand it off to these kids, you know they're conscious."

QUESTION TWO: Since you're all scientists—it is my contention that if we make it easy for each one of us to do the right thing, we will succeed at this. Are you working on making it easier for people to do the right thing? I'd like your feedback on the easy quotient in our future. How does that work into the betterment of our world? It has to be easy to do the right thing. It has to be easy to think about the consequences of traffic, lots of traffic, of combustion engines. If we make it easy for people to do the right thing, I believe they will, because we as humans, don't we always kind of default to easy?

J DESILVA: That's a tough question. I don't think it's going to necessarily be easy, but I hear what you're saying. I think it's got to start with
equity and equal opportunity if you have populations of folks who are
just barely surviving. I see this when I travel to Africa. There's a lot of
exploitation of forest resources in the production of charcoal that is
changing the habitats where the chimpanzees are living, resulting in the
[dramatic] depletion of their numbers. And yet the people there need to
eat and they're going to do what's in their own best interest rather than
thinking of something globally at that moment, and I don't blame them.
I'd be making the same decision if it [were] my family, as well. I think
it's got to start with equity and equal opportunity around the globe, and
that's when folks can start to make those kinds of decisions, because they
have the luxury to do so.

D GRINSPOON: I think that's a great point. Another example: is it's
been said, wisely I think, recently by a few people that if you really want
to fight climate change and overpopulation globally, then contribute to
the education of girls in poorer countries. Because you empower girls,
you empower women to make reproductive choices and other choices
and population growth goes down, because people are not just desperate
and they're making more conscious choices.

Again, it's another example of nobody wants to destroy the world, but
people are going to do what they need to do, or what they feel they need
to do, to survive. If you create those conditions where there's not a conflict between those two, and there are ways we know of that we can. And
some of the news is good. The population projections later in this century all show population peaking and then starting to turn over. It peaks
at ten billion people, which is a little frightening, but it turns over for the
right reason. Not because of an increase in the death rate, but because of
a decrease in the birth rate, because extreme poverty is declining. There
are some of these trends that, if the wind's blowing in the right direction,
we can try to accentuate.

M GLEISER: Or, to play devil's advocate, you could think like JFK,
we are going to the moon and do all these other things because it is
hard. And honestly, at the level of individual agency, to change a life of
habits because you have a purpose [that] is bigger than what you need

immediately is really hard, and it takes, as you said, being able to do it and having the moral attitude and the means to actually make that choice. That's not always very easy. It's actually super hard.

QUESTION THREE: Do you think humans are still evolving from a biological, genetic standpoint, or has human nature reached a point where it becomes static?

J DESILVA: Many folks have argued that once humans became a cultural animal that we had this cultural buffer on natural selection and were no longer biologically evolving. [If] that would have gone back into the Upper Paleolithic, you're talking about fifty thousand years ago. But geneticists have found that there are all these major changes that have happened in the human population in just the last ten thousand years. Changes having to do with eye coloration—increasing the number of individuals with blue eyes, for instance. Changes in the number of genes you have for digesting starches, which would have been really beneficial with the advent of agriculture. We can measure the changes in the frequency of individuals who are born without a third molar. For thirty-five million years, our lineage, going back to the common ancestor with monkeys, had three molars in each quadrant of their mouth. A wisdom tooth was part of our dental arcade and we're beginning to lose it. It's happening in part because not everyone on the planet has access to dental care. We may say, "It's just a toothache." But in some places, that tooth infection can be fatal. When you have that many people on the planet still without access to healthcare and without access to dental care, then of course they're still subject to natural selection. Large swaths of the population are still under the influence of natural selection. Our population is still changing. We're still evolving.

D GRINSPOON: Can I be contrary, and can I say that I feel like the examples you gave of "yes," in a way are a resounding "no," in that none of them are examples of evolving in any sense that is going to change our ability to face the problems that we are facing. I mean it's interesting. We are changing genetically, and so I'm not saying it's false, this statement that we're not. But in terms of the people [who] say it's been outstripped by cultural change and technological change, in terms of our actual manifestation as a species on the planet, it's hard to argue that changes in our teeth and our ability to digest starches is really going to affect our ability to survive at this point.

J DESILVA: And it's hard to predict, right? The duplication of the starch genes—we couldn't have predicted and yet looking back on it, the advent of agriculture is what was the cause of that. The next century, the next two centuries, the changes we're going to face on this planet, those new selection pressures on seven and a half billion people? There's now more variation than ever and that's the raw material natural selection works on. The ingredients are there. We just can't predict what's going to result.

M GLEISER: There is a completely different dimension to your question, though, which is that we are changing big time because of our connection with technology. In a very short time scale of decades, we're not going to be what we are right now, just as we are already different from what we were twenty, thirty years ago. To give you a quick example—our relationship with our cell phones. The vast majority of people on this planet [have] cell phones. And these cell phones are not just interesting little things. They are really digital extensions of the self. If you look at different cell phones, they're all going to have apps. And we all have some similar apps, but each cell phone will have a very unique collection of apps because that is a person's digital fingerprint: an extension of who we are already projected into a machine that is able to take you into other places, allowing you to transcend the flesh and take you across the world in a flash. This, to me, is really where it's going, a blending of humans with machines that is already happening. It is irreversible, and it's going to be deeply transformative.

In fact, the transhumanism movement is taking this to the very extreme, to the point where we will become something very different indeed. And we're talking decades, not centuries and millennia.

QUESTION FOUR: As we're talking about transformation of our culture in order to deal with a threat to our species, I thought this was a really fascinating way of presenting it and you all are academics. Academia is normally about the transmission of information, but I've heard it said that actually, we're far more changed by stories than by the presentation of information. I'm curious to hear your thoughts about how we use that understanding that we need to use storytelling to change this incredibly difficult change that we're seeking for our survival.

T HUSAIN: I don't know how we will do it planet-wide or throughout every academy, but I do see an increased emphasis on science communication, and that's one way of inducing a change. If academics know that it is part of the expectation, in order to get grants or to get tenure, to belong to the system, many more will make an effort to engage the public. And once you do that, you immediately find that no one's going to turn up if you just present them with facts. You have to find some way of making the facts emotionally relevant. Stories are the most generic and the easiest way, but there are others.

The point is to communicate why something matters, so people have a reason to remember what you say and that they begin to see why you do what you do. I think the basic question every audience has for the scientist is why should I care? It's great you're doing this amazing research in your ivory tower, but what does that mean for me? And any effective answer to that question will always have an emotional component—something that tugs at the inside. It's not enough to appeal to intellect alone.

That's one of the reasons we're having so much trouble with climate change. You talk to perfectly reasonable people, tell them all the facts, they listen, but then they say, "OK. But we're still not going to do this" or "I don't agree with you," or "I don't believe in science," which, personally speaking, drives me insane. But, as I keep reminding myself, it's usually not the facts they're arguing, it's that they don't see how and why those facts matter to them.

I think many of us who do outreach realize that we need to start by [establishing] that vital connection. To me, that's far more important than imparting a specific amount of information. In most cases, the relevant facts are so readily available, anyone could look them up. So, if people walk away from a dialogue with a scientist, and they don't remember the facts, I have zero trouble with that. I just want them to [have] a sense of, "Oh, that's why this issue matters, and here's how they're trying to address it."

But there's definitely a learning curve here because academics aren't traditionally encouraged to establish an emotional rapport with audiences. We're trained to sound impartial and objective and completely removed from the situation. I feel adopting that familiar, academic tone with a lay audience is extremely counterproductive, which is why I've become so passionate about science communication being much more nuanced than it currently is. We need to get to where scientists aren't just throwing facts at people, they're sharing what they do, why it matters, and why you should maybe consider coming along for the ride. It's a slow change, but I do think that shift is happening.

M GLEISER: I think that's also why the humanities have a huge role to play. They are the consummate storytellers, through fiction, through artistic creativity. They talk about what is good, what is just, the reason to be human in a completely different dimension. I think we really need to bring these two (the sciences and the humanities) together in order to create this new narrative. It can't just be about the science, because that will push people away. It has to be about the science and about the human nature of the people [who] make that science. And it has to be about why it is important for humanity as a whole to pay attention to these stories, because they are about our collective future.

WHERE DO WE GO FROM HERE?

O ne of the most gratifying aspects of our experiment to bring the engagement of the sciences and the humanities to the public sphere was the extremely supportive feedback from the audience. In our post-event surveys, the question "Should the sciences and the humanities talk to one another?" consistently received a resounding approval margin that varied between 95 and 99 percent. Without a doubt, the general public (of course, self-selected, given that they chose to come to the events) feels that this is an essential dialogue and wants to take an active part in it. They want more events like these and hope that the discussion will be broadened.

Having grown up in a country where the concept of liberal arts education is alien to schools at all levels, I only came to appreciate its importance when I joined the faculty of an American university. In most of the world, students choose their careers too early, often even before joining a university. In many schools, the sciences and the humanities are housed in buildings that are miles apart. There is no cross-disciplinary interaction. Students are served a highly technical and specialized education, becoming, not surprisingly, highly specialized technicians with a poor grasp of disciplines other than their own. As an anecdote from within my own field of physics, when a puzzled undergraduate asks questions about the meaning and philosophical interpretation of quantum physics—the physics of very small objects from molecules to elementary particles largely responsible for the digital revolution—they are usually told to "shut up and calculate!" and are strongly advised not to dare thinking about the philosophical implications of these equations.[1] The further scientists and humanists progress along their course of study, the narrower their research focus tends to become. An educational model that praises hedgehogs over foxes produces efficient professionals to be sure, but professionals with a limited scope of how their jobs impact other sectors of knowledge or society, or how different global trends would necessitate a strategical rethinking of their own targets and goals.[2] When your professional training is built within high walls, it becomes quite difficult to see what lies beyond. Like the hedgehog, all you can do is dig deeper. Educational institutions face the difficult challenge of training professionals who dominate the technical aspects of a field (we do need experts!) while also having an understanding of how their fields relate to others. Perhaps we should take a fresh look at Plato's educational training for philosopher-kings as spelled out in *The Republic*. Humanity needs wise leaders and citizens, well-versed in different ways of knowing. The sciences and the humanities need each other, and we need them both.

A good starting point would be a deep rethinking of science education. The teaching of science needs to move beyond what we could call the conquering mode: a science class is typically about the final results— the laws, the data, the equations—and not about the difficulties of the process, the failures and the challenges that humanize science. This dehumanizing approach to science teaching works as a cleaver, splitting students and the public into two distinct groups: those who embrace it and those who shun it. One of its consequences, as we see in movies and books, is the widespread stereotype of the nerdy cold scientist, focused on the research and not much else, more like a reasoning machine than a human being. This image, obviously totally false, must go. Not only to restore a level of trust in scientists and to value their expertise, but also because science has an enormous impact on society and on our collective future. When it comes to life choices—vaccines, nuclear weapons, genetic engineering, global warming, AI—why should the general public trust the pronouncements of such a specialist? As a timely example, we can see why so many people trust Dr. Anthony Fauci, chief medical advisor to President Biden. He is seen as a human before he is seen as a specialist.

Over the past two centuries, and largely influenced by the profound and immediate impact of technological applications of scientific thinking in industry and society, the teaching of science was mostly reduced to the instruction of technicians, a specialized guild focused on very specific tasks. We became extraordinarily efficient at handling abstruse mathematics and computer programming, of modeling specific systems and handling laboratory demands within narrow subdisciplines: plasma physics, condensed matter physics, high-energy physics, astrophysics, and so on. The walls erected between the sciences and the humanities after the Enlightenment have multiplied into walls erected between the countless subdisciplines within each scientific field, from physics and chemistry to biology and computer science. Reductionism took over education and we lost sight of the whole.

But the whole is where we exist, and our decisions now will affect our collective destiny. We debated several topics that, in the wrong hands, political or corporate, could easily turn into severe existential risks for humanity. What we choose to do with AI or genetic research, the moral boundaries that we impose or fail to impose, will surely affect the very definition of what it means to be human. To plow ahead without considering the ethical consequences of our research, the potentially devastating social changes it can promote—including a catastrophic increase in social disparity and inequity—is at the very least reckless, if not self-destructive. We need a new way to think about our connection to technology, to the marketplace, to the environment, that hinges on how we see the relationship between the sciences and the humanities. It is irresponsible to think that science alone will have all the answers, especially given what we have learned from history. Science is never alone, as it responds to hierarchies of power and cultural contexts. We address more questions that we can answer than questions that we should answer, for those are usually tougher. To address them, we need to think of knowledge in a more inclusive, complementary fashion. From what we have learned from these debates, however, the only path toward survivability is one of cohesion, even if it looks like a distant dream, especially given the current polarization of society. Cohesion is distinct from unity, in that different factions can still move together toward a single goal. A country's army combines people with different opinions and life histories, connected by a cause that transcends their divisions. While we crucify ourselves through relentless tribal disputes and hatred, nature plows ahead, inexorably responding to our insults with mechanic indifference. The simple truth so few are willing to accept is that nature doesn't care about us, but we need to care about nature. We need to develop a kind of environmental herd immunity, a deep rethinking of our relationship with the natural world and each other, that begins with a transformative change at the individual level. Each of us holds a piece of our collective future in our hands. If we pull in opposite directions, we move nowhere. Let us choose wisely how to proceed.

⸺ ◈◈◈ ⸺

When the Institute for Cross-Disciplinary Engagement's assistant director Amy Flockton and I set forth our planning for the public dialogues, we had no expectation of being comprehensive. Our goal was to create an environment where a small sample of difficult questions, questions that have an impact on our inner or outer lives, could be addressed in the open, perhaps as a small spark to help ignite a growing fire. It is our hope that the current volume illustrates that this conversation is not only possible, but urgent, and it must be expanded within academia and in the public sphere.

NOTES

1. This actually happened to me when I was a graduate student in England. I approached the venerable physicist John Bell, asking if he would be interested in advising me on a Ph.D. thesis on the interpretation of quantum physics. His answer, although kind, was direct: "Don't waste your time doing this until you are famous. No one will pay attention to you."

2. The Ancient Greek poet Archilochus famously said: "A fox knows many things, but a hedgehog knows one big thing." The idea was picked up many times later, most famously in a 1953 essay by Isaiah Berlin, "The Hedgehog and the Fox," where he used it to classify ("not very seriously," he later claimed) writers and thinkers into two kinds.

ACKNOWLEDGMENTS

This volume would not have been possible without the presence and the wisdom of all of those who shared the stage with me during these past few years. I am profoundly grateful to you all, for your eagerness to join me in these conversations, for your intellectual openness, and for joining me in the essential mission of narrowing the gap between the sciences and the humanities. There is much work to be done, for sure, but we have certainly taken a step forward, as have other universities and foundations across the country that have started similar initiatives in the past few years. The momentum is growing, and we need all the resources we can muster.

A very special thanks to the John Templeton Foundation and, in particular, to Heather Dill Templeton, for her enthusiastic support throughout the years. To Chris Levenick, the foundation's director of public engagement, who continually steered our work at the Institute for Cross-Disciplinary Engagement in the right direction, and to Jessica Hunt, whose help with grant logistics was essential. I also would like to thank the indefatigable Mary-Ann Myers and Ayako Fukui for their initial help and encouragement in getting us going.

My most heartfelt thanks goes to Amy Flockton, the Institute's assistant director and its heart and soul, and now a dear friend. Without Amy, the Institute wouldn't have done half the things it did, and those would certainly have been less effective. Amy played an essential role in setting up the public dialogues assembled in this volume (and the others not included here), in getting all the often complex logistics going, and in obtaining the permissions from our speakers. Most importantly, Amy also assembled the transcripts in this volume and edited them into readable form. Without her help and encouragement this book wouldn't exist.

I'd also like to thank my colleagues at Dartmouth for their support during the years, in particular the president, Phil Hanlon, a strong advocate for the engagement across the disciplines. A heartfelt thanks to Wendy Lochner at Columbia University Press, for championing this volume and for her continuous enthusiasm.

CONTRIBUTORS

Ed Boyden, one of the leading neuroscientists working at the cutting edge of this topic, is the head of the Synthetic Neurobiology Group and an associate professor of biological engineering and brain and cognitive sciences at the MIT Media Lab and at the McGovern Institute for Brain Research. Ed is really trying to make headway in some of the most difficult questions concerning the notion of how the brain engenders consciousness, what that means, and whether we can reproduce this process (if that's what it is) artificially or not.

Jimena Canales grew up in Mexico, then came to the United States, where she did a master's and a PhD at Harvard in history of science. She stayed at Harvard, teaching for many years, and then she moved to the University of Illinois as the Thomas Siebel Professor of the History of Science. She recently returned back to Cambridge, Massachusetts, to dedicate herself to writing. She's the author of many technical articles and journals, and also of two books, *A Tenth of a Second: A History* and *The Physicist and the Philosopher*.

The Physicist and the Philosopher has received much praise. It was voted one of the best science books of 2015 by NPR's *Science Friday* and by Maria Popova's *Brain Pickings*. It was listed as a Top Read by the *Independent* and named one of the Books of the Year by the *Tablet*. Jimena's first book, *A Tenth of a Second*, was listed as one of the *Guardian's* Top

10 Books About Time. She has also written many essays for the *New Yorker*, the *Atlantic*, *Wired*, the BBC, and other venues.

Sean Carroll is a world-renowned theoretical physicist, specializing in cosmology, something called quantum field theory, and general relativity. He is currently a research professor at the California Institute of Technology in Pasadena. Sean has written many articles and essays for *Nature*, the *New York Times*, *New Scientist*, etc. He has authored four popular science books, the latest being *Something Deeply Hidden*, on the interpretation of quantum mechanics, a very controversial subject. He has appeared on many TV shows about cosmology, and science in general, on the History Channel, on *Through the Worm Hole with Morgan Freeman*, and on the *Colbert Report*, which he survived to tell the tale of, not a small feat. His research focuses on the physics of the very early universe and the nature of the mysterious dark energy. His PhD is from Harvard and he was a post-doctoral fellow at MIT, and the Kavli Institute for Theoretical Physics in Santa Barbara, a few years after I was there. He was a professor at the University of Chicago before becoming a professor at Cal Tech.

David Chalmers is a university professor of philosophy and neural science and co-director of the Center for Mind, Brain, and Consciousness at New York University. He also has an appointment at the Australia National University, where he used to direct the Centre for Consciousness. He works on the philosophy of mind and language, and on artificial intelligence and cognitive science. He's well known for his formulation of the *hard problem of consciousness*, which is the problem of explaining how and why we have first-person experiences. Essentially, we could phrase it like this: Why does the feeling that accompanies the awareness of sensory information exist at all? He's the author of *The Conscious Mind* and, more recently, *The Character of Consciousness*, as well as many philosophical essays (I have to add that he's also the lead singer of the Zombie Blues band).

Patricia Churchland is Professor Emerita of Philosophy at the University of California at San Diego, where she is an adjunct professor at the Salk Institute. Her research focuses on the interface between neuroscience and philosophy. Her most recent book is *Conscience, The Origins of*

Moral Intuition. She is the author of the groundbreaking books *Touching A Nerve* and *Neurophilosophy*, and she co-authored, with Terry Sejnowski, *The Computational Brain*. Her book, *Brain Trust: What Neuroscience Tells Us About Morality*, received the PROSE prize for science. She has been president of the American Philosophical Association and the Society for Philosophy and Psychology. She won the MacArthur [Fellowship] in 1991 and the Rossi Prize for neuroscience in 2008. She was chair of the USD Philosophy Department from 2000 to 2007.

Antonio Damasio is University Professor, David Dornsife Chair in Neuroscience, and Professor of Psychology, Philosophy, and Neurology at the University of Southern California, where he also directs the Brain and Creativity Institute. He has made seminal contributions to the understanding of brain processes, underlying emotions, feelings, decision-making, and consciousness. He has authored four books, most recently *Strange Order of Things: Life, Feeling and the Making of Cultures*. And he has an astounding 183,100 citations for his academic work.

Paul Davies is a world-renowned theoretical physicist. He works in cosmology, in astrobiology, and also in astrophysics and the theory of black holes. He has written many books, and he's also a Regents Professor here at ASU and the director of the Beyond Center for Fundamental Concepts in Science.

He has written three books explicitly on time. One of them, a more technical one, from 1974, is called *The Physics of Time Asymmetry*. And then two others addressed at a more general audience, *About Time* from 1995, and, the one that I would love to know the answer to, *How to Build a Time Machine* from 2001. Paul may not even be here right now—it may be his future self, visiting us.

He's won many awards, including the Templeton Prize, and also the [William Thomson, Lord] Kelvin Medal and the Faraday Prize, which are given to people who have a true impact on the public understanding of science.

Jerry DeSilva is an associate professor of Anthropology at Dartmouth College. He is a paleoanthropologist specializing in the locomotion of the first apes, hominoids, and early human ancestors, hominins. His particular anatomical expertise, the human foot and ankle, has contributed to our understanding of the origins and evolution of upright walking in the human lineage. He has studied wild chimpanzees in west Uganda and early human fossils in museums throughout eastern and south Africa. From 1998 to 2003, Jeremy worked as an educator at the Boston Museum of Science and continues to be passionate about science education.

Rebecca Goldstein is a very acclaimed philosopher and novelist. She's the recipient of numerous prizes for her fiction and scholarship, including a Guggenheim Fellowship and a MacArthur [Fellowship]. In 2012, she was named a Humanist of the Year by the American Humanist Association. In 2015, she received the National Humanities medal from President Obama. She is the author of ten books, including *Plato at the Googleplex: Why Philosophy Won't Go Away* and a wonderful novel called *Thirty-Six Arguments for the Existence of God: A Work of Fiction*. She is currently a visiting professor at the New York University as well.

David Grinspoon is currently at Dartmouth as an ICE [Institute of Civil Engineers] Fellow. He's an astrobiologist. For those of you who do not know what that means, it means that it's OK now for scientists to think about extraterrestrial life. You actually get grants to do it. It's a wonderful time. He is an award-winning science communicator and a prize-winning author. His newest book is called *Chasing New Horizons: Inside the Epic First Mission to Pluto*, co-authored with Alan Stern. He's a senior scientist at the Planetary Science Institute in Washington and an Adjunct Professor of Astrophysical and Planetary Science at the University of Colorado. He's involved with several interplanetary spacecraft missions for NASA, the European Space Agency, and the Japanese Space Agency. His book *Earth in Human Hands* was named a Best Science Book of 2016 by NPR's *Science Friday*. His previous book, *Lonely Planets: The Natural Philosophy of Alien Life*, won the PEN Center USA Literary Award for Nonfiction.

Tasneem Zehra Husain came to writing via theoretical physics. She received her PhD from Stockholm University in Sweden and pursued post-doctoral research at Harvard, before returning to her native Pakistan as a founding faculty member at an elite school of science and engineering. She's particularly passionate about the need for a more nuanced, more human paradigm in science writing, one that is truer to both the process and the spirit of the endeavor. Tasneem has conducted several writing workshops for scientists, including an ongoing series at CERN. Her writing has appeared in *Nautilus*, as well as various anthologies of science writing for both adults and children. She's a regular columnist for 3quarksdaily.com and the author of the popular science novel *Only the Longest Threads*. For over a decade Tasneem has been actively involved in outreach. She has worked with K–12 teachers, high school students and government officials, both in the United States and abroad, and is a frequent speaker at book festivals, science festivals, writing conferences, and physics conferences. She sees humans as storytelling creatures.

Elizabeth Kolbert has been a staff writer at the *New Yorker* since 1999. Her three-part series on global warming, "The Climate of Man," won the 2006 National Magazine Award for Public Interest, among other honors. She received a Lannan Literary Fellowship in 2006, a Heinz Award in 2010, and won the 2010 National Magazine Award for Reviews and Criticism. She is the author of *The Prophet of Love: And Other Tales of Power and Deceit*, *Field Notes from a Catastrophe*, *The Sixth Extinction*, for which she won the 2015 Pulitzer Prize for General Nonfiction, and most recently, *Under a White Sky: The Nature of the Future*.

Alan Lightman is a physicist, a novelist and an essayist—the first at MIT to receive dual appointments in the sciences and the humanities. I'm very jealous. He's the author of five novels, two collections of essays, a booklength narrative poem, and many nonfiction books on science. His novel *Einstein's Dreams* was an international bestseller and published in thirty languages. His most recent books include *Screening Room: A Memoir of the South*, one of the *Washington Post*'s 2015 best books of the year, and *The Accidental Universe*. His latest book is *Searching for Stars on an Island in Maine*, an inspiring and personal meditation on science and spirituality. On top of all this, Alan has received the gold medal for humanitarian service to Cambodia awarded by the government of Cambodia, celebrating the work of his Harpswell Foundation, a nonprofit with a mission to equip young women in Southeast Asia with leadership and critical thinking skills.

Siddhartha Mukherjee is a pioneering physician, oncologist, and author. A very influential voice in the scientific community, he's best known for his books *The Emperor of All Maladies: A Biography of Cancer*, which earned him the 2011 Pulitzer Prize, and *The Gene: An Intimate History*, which was recognized by the *Washington Post* and the *New York Times* as one of the most influential books of 2016. *The Emperor of All Maladies* was adapted into a documentary by Ken Burns and included among *Time* magazine's All-Time100 Best Nonfiction Books.

Mark O'Connell is a journalist and humanist who has addressed these questions in a book that explores and meditates on transhumanism and what it means to us as individuals and as a species, *To Be a Machine: Adventures Among Cyborgs, Utopians, Hackers and the Futurists Solving the Modest Problem of Death*. In 2018, the book won the UK's very prestigious Wellcome Book Prize.

Jill Tarter holds the Bernard M. Oliver Chair for SETI (Search for Extraterrestrial Intelligence) Research at the SETI Institute in Mountain View, California, and serves as a member of the board there. Tarter received her Bachelor of Engineering Physics at Cornell University and a master's and a PhD in Astronomy at the University of California. She has spent the majority of her professional career attempting to answer the age-old and essential question, "Are we alone?"—the question we all want to know the answer to—by searching for evidence of technological civilizations beyond Earth.

Since the termination of funding for NASA's SETI program in 1993, she has served in leadership roles to design the Allen Telescope Array and to secure private funding to continue the exploratory science of SETI. If you want to be participatory on this, fund SETI. Many people are now familiar with her work, as portrayed by Jodie Foster in the movie *Contact* but, more to the point, there is a recently published authorized biography of Jill Tarter by science writer Sarah Scoles called *Making Contact*.

B. Alan Wallace is an expert in Tibetan Buddhism, having been ordained as a monk by the Dalai Lama himself, after studying with him for about fourteen years. His many books discuss Eastern, Western, scientific, philosophical, and contemplative modes of inquiry, often focusing on relationships between the sciences and Buddhism.

Alan is the founder of the Santa Barbara Institute for Consciousness Studies and is very active as a lecturer and instructor all over the world. He spends only three months a year in Santa Barbara. The rest of the time he travels, presenting workshops in places from Santa Barbara to Tuscany and beyond. He has a bachelor's degree in physics from Amherst College, and a PhD in religious studies from Stanford. He is definitely a person to listen to when we want to explore relationships between the sciences and spirituality, in particular from the Tibetan Buddhist tradition.

INDEX

CPSIA information can be obtained
at www.ICGtesting.com
Printed in the USA
JSHW031530260622
27275JS00002B/2